Lecture Notes in Mobility

Series editor

Gereon Meyer, Berlin, Germany

More information about this series at http://www.springer.com/series/11573

Beate Müller · Gereon Meyer
Editors

Electric Vehicle Systems Architecture and Standardization Needs

Reports of the PPP European Green Vehicles Initiative

Springer

Editors
Beate Müller
VDI/VDE Innovation + Technik GmbH
Berlin
Germany

Gereon Meyer
VDI/VDE Innovation + Technik GmbH
Berlin
Germany

ISSN 2196-5544 ISSN 2196-5552 (electronic)
Lecture Notes in Mobility
ISBN 978-3-319-13655-4 ISBN 978-3-319-13656-1 (eBook)
DOI 10.1007/978-3-319-13656-1

Library of Congress Control Number: 2015930734

Springer Cham Heidelberg New York Dordrecht London

Printed on acid-free paper

Springer International Publishing AG Switzerland is part of Springer Science+Business Media (www.springer.com)

Foreword

Electronic components and ICT systems are ubiquitous and bring a high value to today's vehicles. Steady electrification is conferring them an even more defining role and value share in the transportation means of the future with ICT being instrumental for most building blocks of an electric car.

Complementing materials innovation, ICT-enabled components and services are significantly reducing the energy demand of vehicles and improving the safety of the transport system, directly leading to a large societal impact. Beyond that, they make driving more comfortable.

DG CONNECT is the driver of the ICT pillar of the European Green Vehicle Initiative PPP (EGVI PPP). More than 30 R&D projects with over 110 million € funding were launched since the first call in 2009. Several large-scale automotive ICT projects are also supported under the Joint Technology Initiatives ENIAC and ARTEMIS. All these projects are now delivering tangible results. Research on electronic/electronical architectures has been a ground-breaking topic with a significant industrial impact. Projects like eFuture or OpEneR are showcasing the benefits of cross-border cooperation along the value-chain granting the European industry a competitive edge.

The EGVI has grown from a recovery programme for crisis-ridden sectors into a strategic longer-term consolidated instrument with a strongly committed and active community. The contractual arrangement of the European Green Vehicles PPP was signed on 17 December 2013 by the Commission and representatives of the sector, showing the long-term commitment of the European Union to financing R&D&I in the sector under Horizon 2020.

A further substantial opportunity to support collaborative automotive R&D&I under H2020 is available through the JTI Electronic Components and Systems for European Leadership (ECSEL) launched in early 2014. Automotive stakeholders and platforms are encouraged to actively participate.

Identifying future European policy and support priorities is a non-trivial task. It needs a close cooperation of all actors in a rapidly evolving landscape with changing paradigms. Standardised architectures and by-wire technologies have the potential to pave the way for European automotive USP. The "programmable car" enabling functional integration may generate competitive vehicle performance and added value. Autonomous driving made it from private research labs to collaborative programmes and enjoys strong media coverage. Big data and data security are also considered key for the smart, connected vehicles of tomorrow.

With strong public and private support, the first fully electric vehicles from series production have recently rolled out, roughly one century after the invention of the electric car. Old and new value-chain players could now grasp this window opportunity and momentum to foster positions and innovate further.

Khalil Rouhana
Director of Directorate A, Components and Systems
Directorate-General for Communications Networks
Content and Technology (DG CONNECT), European Commission

Disclaimer The views expressed in this note are the sole responsibility of the author and in no way represent the view of the European Commission and its services.

Preface

The objectives of sustainable road mobility, i.e. energy efficiency, climate protection and zero emissions, imply a paradigm shift in the concept of the automobile regarding its architecture, design, materials and propulsion technology. The electric vehicle (EV) is seen as the most viable option. However, it is still facing a multitude of challenges in terms of product maturity and user acceptance. Moreover, the growing market share of EVs inevitably leads to a renovation of the classical automotive value chain and will result in a shift in the creation of added value in the supply chain.

The Coordination and Support Action "Smart Electric Vehicle Value Chains (Smart EV-VC)" funded in the Seventh European Framework Programme, analysed these novel smart EV supply chains and possible supporting measures for their strengthening in Europe. This analysis was based on the identification of the unique selling propositions (USP) of the European smart EV which should be served by the adapted value chains. These USPs have been found to be: affordability, smartness and connectivity, adaptation to mobility needs and use patterns and safety and reliability. On technology level, most of these USPs are related to overcoming today's drawbacks of EV batteries that lack energy density, lifetime and affordability.

In a smart approach range extension may be reached in an intelligent way by enabling battery downsizing through implementing ICT and smart systems and components, since integrating a high degree of electronic control, adaptive capabilities and intelligence to the system may raise energy efficiency significantly. Especially, since in EVs most mechanical control functions can easily be replaced by electronic means and are supported digitally by embedded software, these synergies present a parallel path to innovations in cell technology or use of lightweight materials. Hence, they may greatly support the removal of barriers to the wide implementation of the electric vehicle.

Experience with comparable transitions from mechanically via electrically to electronically and digitally controlled systems (e.g. from the typewriter to the computer) tells that a significant cost reduction can be achieved when a complete redesign of the platform is undertaken. Hence, for the future generation EVs that conform to the aforementioned USPs, a real paradigm shift can be foreseen: a complete redesign of the electric, electronic and ICT architecture of the fully electric vehicle.

Several research projects of the European Green Vehicles Initiative Public Private Partnership (EGVI PPP) are already addressing topics connected to the USPs and the development of new vehicle architectures and ICT platforms. Some of them were reviewed within a workshop of the EGVI PPP on the topic of electrical and electronic architecture of EVs and EV standardization needs which took place on 23 October 2013 in Brussels. The workshop strived to evaluate the research activities within the EGVI PPP and also to directly gather feedback from the stakeholder groups regarding R&I strategies and funding policies. The scientific talks were complemented by talks on the strategic topics of standardization and support of SMEs. Both topics are important when discussing measures for strengthening the European smart EV value chain. Papers of selected presentations of this workshop are collected in this book.

The EGVI PPP was established as European Green Cars Initiative PPP within the scope of the 7th Framework Programme. In Horizon 2020, the EGVI PPP focuses on energy efficiency and alternative powertrains. Through the duration of the Public Private Partnership in FP7, a close dialogue between the stakeholders of the industry, research institutes and European Commission has been constituted. Among other things, this is expressed in the continuously held expert workshops which are a collaborative activity of the European Commission and the industry platforms European Technology Platform on Smart Systems Integration (EPoSS) and European Road Transport Research Advisory Council (ERTRAC). These workshops were organized by the Coordination Actions "Implementation for Road Transport Electrification" (CAPIRE) and Smart EV-VC.

The aim of this volume of the "Reports of the PPP European Green Vehicles Initiative" is to disseminate the results of the European Green Vehicles Initiative PPP to a wider stakeholder community and to further reinforce the dialogue among the stakeholders as well as with policy makers.

Beate Müller
Gereon Meyer

Contents

Part I
Invited Papers

Current Issues in EV Standardization

Peter Van den Bossche, Noshin Omar, Thierry Coosemans and Joeri Van Mierlo

Abstract In urban traffic, due to their beneficial effect on environment, electrically propelled vehicles are an important factor for improvement of traffic and more particularly for a healthier living environment. The operation of the electrically propelled vehicle is dependent on the availability of efficient electric energy storage devices: the traction batteries, which have to access suitable recharging infrastructures. For all these components, standards are essential for ensuring safety and compatibility. This article gives an overview of current developments in the field of international standardization of electrically propelled vehicles, focusing on two essential matters for electric vehicles: batteries and charging.

Keywords Electric vehicles · Standardization · Charging infrastructure

1 Introduction

The electric vehicle encompassing both automotive and electrical technologies, standardization is not a very straightforward issue. Standardization, on a global level, being mainly dealt with by two institutions: the *International Electrotechnical Commission* (IEC), and the *International Organization for Standardization* (ISO), the question arose which standardization body would have the main responsibility for electric vehicle standards.

P. Van den Bossche (✉) · N. Omar · T. Coosemans · J. Van Mierlo
ETEC, Vrije Universiteit Brussel, Pleinlaan 2, 1050 Brussel, Belgium
e-mail: pvdbos@vub.ac.be

N. Omar
e-mail: noshomar@vub.ac.be

T. Coosemans
e-mail: thierry.coosemans@vub.ac.be

J. Van Mierlo
e-mail: jvmierlo@vub.ac.be

B. Müller and G. Meyer (eds.), *Electric Vehicle Systems Architecture and Standardization Needs*, Lecture Notes in Mobility,
DOI 10.1007/978-3-319-13656-1_1

Table 1 Basic division of work IEC/ISO

ISO	IEC
Work related to the electric vehicle as a whole	Work related to electric components and electric supply infrastructure

One can discern a fundamentally different approach taken towards the concept of standardization in the automotive and the electrotechnical world. There is a different "standardization culture", the origin of which can be traced back to historical reasons.

This difference is further reflected in the constitution of the technical committees and their working groups which deal with electric vehicle standardization in respectively IEC and ISO. In the IEC committees many of the delegated experts are electricians or component manufacturers, whileas in ISO there is a much stronger input from vehicle manufacturers. During the years, there have been considerable discussions between the two groups as to the division of the work, leading to a consensus defining the specific compentences of the respective committees, as shown in Table 1.

Within Europe, CENELEC and CEN operate as the pendants of IEC and ISO. Both have been active in electric vehicle standardization in the 1990s, through their technical committees CENELEC TC69X and CeN TC301. Initially working in parallel to the global standardization work, these committees went dormant around the turn of the century, but TC69X was reactivated in 2011, with the aim of expediting the European adoption of IEC TC69 documents.

2 Battery Standards

The standardization of batteries for electric vehicle traction presents several aspects, including performance, dimensions and safety.

2.1 Battery Performance Standards

The aim of battery performance standards is to assess the operational characteristics of the battery as a "RESS": rechargeable energy storage system. RESS need to provide both energy (for driving range) and power (for acceleration), and are characterized by specific energy (Wh/kg) and specific power (W/kg), both values being illustrated in the Ragone diagram (Fig. 1). For determining the actual performances of the battery, suitable test cycles are needed which reflect the actual use of the battery in the vehicle.

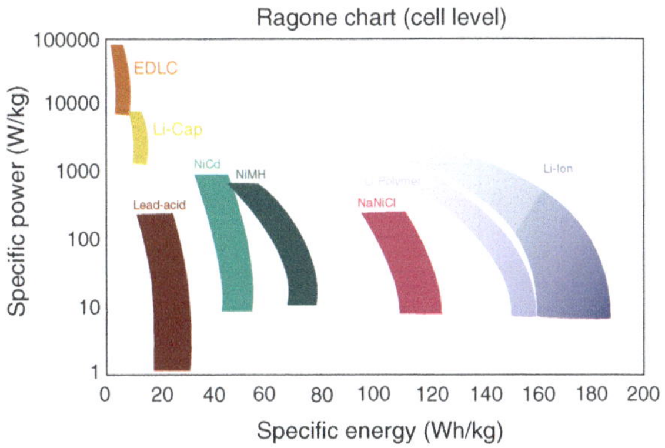

Fig. 1 Ragone diagram

Traditional test cycles such as used for lead-acid industrial traction batteries [1] are based on constant current cycling and are not suited for electric vehicle applications, where the batteries are discharged in a much more dynamic way, and where regenerative braking is used.

New challenges for standardization included both the emergence of new battery chemistries besides lead-acid (alkaline nickel batteries, and of course lithium-ion) and the development of new applications such as hybrid vehicles where the batteries are being used in a different way more based on power storage.

For non-lithium technologies, the IEC61982 "Secondary batteries (except lithium) for the propulsion of electric road vehicles—Performance and endurance tests" [2] describes dynamic power performance tests featuring acceleration, cruising and regenerative braking (Fig. 2).

For lithium traction batteries, standardization has been addressed both by ISO and IEC, focusing respectively on the battery system as vehicle component, and the individual battery cells, leading to the standards ISO12405-1 [3] for power-oriented batteries, ISO12405-2 [4] for energy-oriented batteries, and IEC62660-1 [5] for individual cells.

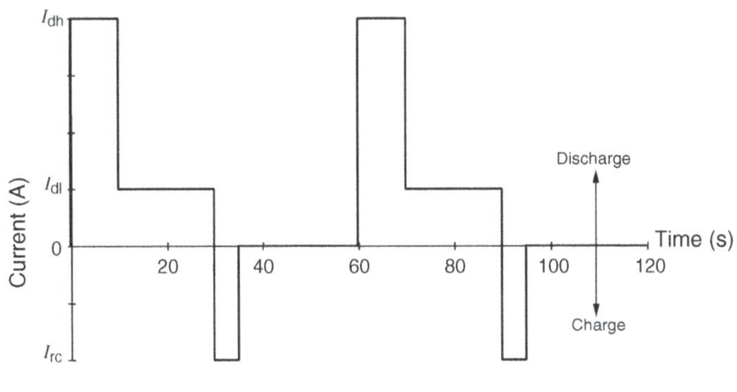

Fig. 2 Dynamic power performance test micro-cycle [2]

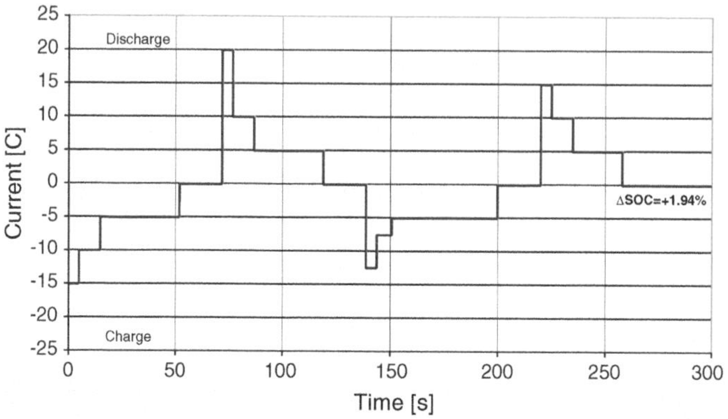

Fig. 3 Hybrid micro-cycle [3]

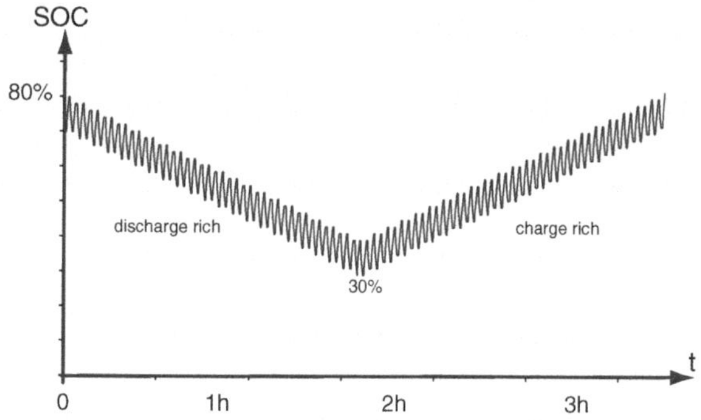

Fig. 4 State of charge evolution for hybrid test [3]

The micro-cycles for hybrid operation are charge-rich (Fig. 3) or discharge rich, and are performed in a limited state-of-charge window (Fig. 4) as is customary for hybrid operation.

The power oriented operation of the battery also necessitates test cycles for pulse power and internal resistance, described on both battery and cell level in the respective ISO and IEC standards. An example is given in Fig. 5.

New developments may be necessary in view of the exchange of batteries between vehicles and the deployment of "second life" vehicle batteries for other applications such as stationary energy storage for grid support. Such applications need a means to accurately estimate the "state of health" of a battery, a difficult process as it is highly dependent on the understanding of a battery's chemistry and environment and the evolution of ageing processes. The CEN-CENELEC Focus Group recommends that parameters for state of health should be defined in standards to allow for second life use of batteries [6].

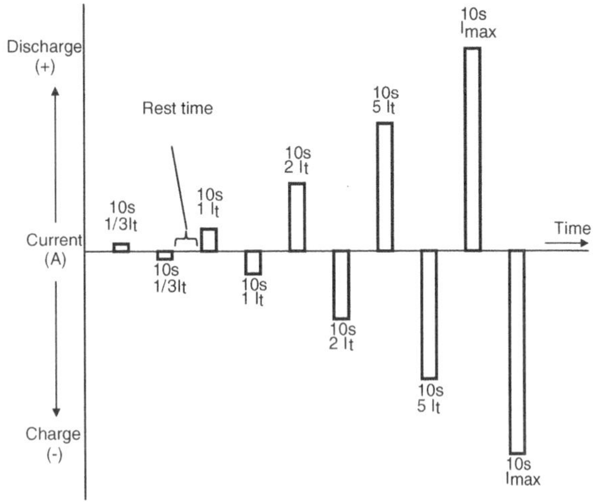

Fig. 5 Pulse power test [5]

2.2 *Battery Safety Standards*

Safe installation of the battery onboard the electric vehicle is treated by ISO6469-1 [7]. Special considerations for post-crash safety, focusing on the risks for emergency personnel, are described in the new ISO6469-4 now under development [8].

For batteries with aqueous electrolyte, such as lead or nickel batteries, hydrogen emission during charging may be a safety hazard, which is treated in IEC62485-3 [9].

Lithium batteries however may present specific hazards due to thermal runaway which may affect some lithium chemistries. This is addressed two-fold in the standards: on one hand, thermal, mechanical and electrical abuse tests are described in the standards ISO12405-1 [3], ISO12405-2 [4] and IEC 62660-2 [10]; on the other hand, pass/fail criteria for these tests are developed in the forthcoming standards ISO12405-3 [11] and IEC62660-3 [12], with ISO and IEC acting respectively on system and cell level.

2.3 *Battery Dimensional Standards*

For mature technologies such as industrial lead-acid batteries, dimensional standards such as IEC60254-2 [13] are well established. For lithium however, the technology is still evolving and it might be stifling to fully standardize dimensions just now. Lithium cells come in various sizes and shapes (cylindrical, prismatical, pouch-format) and various chemistries. In order to provide design guidelines, the Publicly Available Specification (not a full-fledged standard) ISO/IEC16898 [14] was issued, defining designations and markings of cell dimensions, configurations and position of terminals and venting mechanism, which are to be used for design of battery packs. This document contains no specifications for inner construction, cell chemistry, electrical characteristics and any further properties, and neither describes relation between dimensions and capacity of cell as the performance of secondary lithium-ion batteries for vehicle propulsion is still being improved quickly.

3 Charging Standards

3.1 Conductive Charging Standardization

The main reference documents for conductive charging are the IEC61851 family of standards. The first part, dealing with general requirements, was first published in 2001 [15]. Revision was started with the reactivation of IEC TC69 WG4 in 2005, leading to the publication of the second edition in 2010 [16]. Work on the third edition is now ongoing, with CD versions circulating in 2012 [17] and 2013 [18].

Part 21, initially dealing with vehicle requirements, saw its vehicle requirements proper transferred to ISO (as vehicle-related issues are ISO's province) into a new document ISO17409 [19], with Part 21 focusing on EMC issues for charging [20, 21]. Parts 23 and 24, dealing with d.c. charging, were published in 2014 [22, 23]. To cover the specific needs of light electric vehicles, a new work item proposal was adopted in 2012 [24].

3.2 Charging Modes for Conductive Charging

The standard IEC61851-1 defines the so-called *charging modes* for conductive charging [16].

3.2.1 Mode 1 Charging

Mode 1 charging refers to the connection of the EV to the a.c. supply network (mains) utilizing standard (non-dedicated) socket-outlets with currents up to 16 A. The safe operation of a Mode 1 charging point depends on the presence of suitable protections on the supply side: a fuse or circuit-breaker to protect against over-current, a proper earthing connection, and a residual current device (RCD). It is difficult however for the EV driver to easily assess the quality and safety of the electrical installation when plugging in. For this reason, Mode 1 charging is being deprecated except for light vehicles (such as two- and three-wheelers) which can be made as Class II equipment (with double insulation).

3.2.2 Mode 2 Charging

Mode 2 charging connection of the EV to the a.c. supply network (mains) also makes use of standard non-dedicated socket-outlets. It provides however additional protection by adding an *in-cable control box* (ICCB).

Mode 2 is now generally proposed for convenience charging at non-dedicated outlets. The main disadvantage of Mode 2 is that the control box protects the downstream cable and the vehicle, but not the plug itself, whereas the plug is one of the components more liable to be damaged in use.

3.2.3 Mode 3 Charging

Mode 3 charging: involves the direct connection of the EV to the a.c. supply network utilizing dedicated electric vehicle supply equipment. This refers to private or public charging stations. The standard IEC61851-1 [16] mandates control pilot protection between equipment permanently connected to the a.c. supply network and the electric vehicle.

The so-called *control pilot* device has the following functions mandated by the standard:

- verification that the vehicle is properly connected
- continuous verification of the protective earth conductor integrity
- energization and de-energization of the system
- selection of the charging rate (ampacity)

This function is typically performed through an extra conductor in the charging cable assembly, in addition to the phase(s), neutral and earth conductor. Annex A of IEC61851-1 (published as technical specification IEC/TS62763 [25] pending publication of third edition of 61851-1) specifies the control pilot circuit. A control signal (1 kHz PWM, with the duty cycle allowing ampacity control) is sent through the control pilot conductor. When no vehicle is connected to the socket-outlet, the socket is dead; power is delivered only when the plug is correctly inserted and the earth circuit is proved to be sound.

The inherent safety features, as well as the potential for smart grid integration, make Mode 3 the preferred solution for public charging stations as well as for home charging using dedicated outlet [6].

3.2.4 Mode 4 Charging

Mode 4 charging is defined as the indirect connection of the EV to the a.c. supply network (mains) utilizing an off-board charger where the control pilot conductor extends to equipment permanently connected to the a.c. supply.

This pertains to d.c. charging stations, which are mostly used for fast charging. As the charger is located off-board, a communication link is necessary for regulated d.c. charging stations to allow the charger to be informed about the type and state of charge of the battery as to provide it with the right voltage and current.

3.3 Standardization for Fixed Charging Infrastructure

The emergence of fixed charging infrastructure for electric vehicles has also prompted other committees to work on the subject.

Concerning the requirements for charging posts in the public domain, IEC SC17D is preparing IEC61439-7 "Low-voltage switchgear and controlgear assemblies—Part 7: Assemblies for specific applications such as marinas, camping sites, market squares, electric vehicles charging stations", which was circulated as FDIS early 2013 [26]. This document is to be used with the general standard IEC61439-1 [27].

The subject of charging infrastructure was also taken up by IEC TC64, the committee in charge of "electrical installations and protection against electric shock". The general standard IEC60364 "Low voltage electrical installations" is being complemented with a special part dealing with supply of electric vehicles: IEC60364-7-722 [28].

3.4 Wireless Charging

Wireless charging dispenses with the use of cables and connectors. The wireless energy transfer between the vehicle and the charging point can be performed in several ways:

- inductive, through magnetic fields
- capacitive, through electric fields
- microwave, through electromagnetic radiation

The latter two techniques are still in an early experimental stage and any standardization work is still under consideration; significant technological development has taken place however on inductive charging.

The introduction of inductive charging systems has been proposed to allow a considerable improvement of charging safety. The non-conductive energy transfer virtually eliminates all risk of electric shock for the user. Furthermore, the opportunity for automatic connection dispenses with the use of electric cables, thus removing both electrical (handling of power connectors,...) and mechanical (trailing cables,...) hazards which are usually associated with the use of electric vehicle charging equipment.

One type of inductive charging has been introduced and extensively promoted by General Motors in the 1990s. The secondary coils were arranged around a slot in the vehicle, the primary coil being a paddle to be inserted in the slot. This approach, still needing a cable, has been abandoned however.

New developments for wireless charging will make use of automatic systems operating when the car is parked, or even dynamically during driving on an adapted road. The standardization work on the subject has been revived, focusing firstly on safety aspects involved with exposure to magnetic fields in the vicinity of the inductive coils [29].

3.5 Battery Exchange

A particular fast replenishment of the energy on board of the vehicle can be performed by a fast replacement of the battery pack with a freshly charged one. This technology, which has been used in the past for niche applications such as industrial electric vehicles, has now gained new interest for general use. Its implementation however will entail specific standardization problems.

Standardization work on battery exchange has been taken up in 2012 with a new project which will lead to the IEC62840 family of standards [30], covering the safety aspects of the systems which are operated in the battery swap infrastructure premises including: electric, mechanical, structural hazards, fire, risk assessment.

3.6 EMC Issues for Charging

The influence of the extended use of power electronic converters as used in battery chargers will have to be closely followed up in order to avoid potential problems regarding electromagnetic compatibility either in the form of radiated electromagnetic waves or as conducted interference on the interconnecting cables.

EMC is heavily regulated by European directives (2004/108/EC [31] and 2004/104/EC [32] as well as treated in numerous international standards published by IEC, ISO and CISPR.

The EMC constraints for road vehicle have been traditionally focused on radiated EMC. Grid connection of electric vehicles however brought however the problem of conducted EMC and hence the need for new standards.

New standardization work on EMC for electric vehicle charging were undertaken with the revision of IEC61851, where the Part 21 will focus on EMC issues, bringing together all relevant requirements. This document will have two parts, focusing respectively on on-board and off-board charging systems. Both were circulated as CD in 2012 [20, 21].

4 Communication Standards for Charging

4.1 Basic Communication

The communication between the vehicle and the charging post can be developed in several ways, with increasing sophistication. In Mode 1 or Mode 2 charging, where standard non-dedicated socket outlets are used, there is no communication at all. Mode 3 introduces communication through the control pilot function, with ampacity control conveyed by a pulse-width modulation (PWM) signal in the control pilot circuit. This feature presents several operational benefits: the charger can adjust itself to the maximum allowable current that can be delivered by various charging points, and the charging point can control the amount of current absorbed by the charger, in the framework of a smart grid load management or to optimize the tarification of the electric energy.

Mode 4 off-board chargers, which supply a direct current to the vehicle battery, must communicate with the vehicle in order to supply the battery with the correct voltage and current. This is treated in the new standard IEC61851-24 [23], defining the messages of digital/data communication to be used during charging control between off-board d.c. charging system and electric road vehicle.

4.2 High-Level Communication and Grid Management

The development of new concepts such as "smart grid" or "vehicle to grid" has created the need for higher level communication, involving several actors, including both vehicle manufacturers and utilities.

This issue is being addressed by a joint working group uniting ISO TC22 SC3 and IEC TC69, drafting a family of standards called ISO/IEC 15118, to describe the communication between the electric vehicle and the electric vehicle supply equipment (charging post).

The basic document of the ISO/IEC 15118 family cases is part 1 "General information and use-case definition" [33], providing a general overview and a common understanding of aspects influencing the charge process, and contextualizing all envisageable charging processes in so-called "use cases" in order to define communication needs.

Further parts of ISO/IEC15118 describe the technical protocol [34], physical and data link layer requirements [35], as well as test procedures [36, 37] and provisions for wireless communication [38–40].

5 Accessories for Charging

5.1 Generalities

Conductive connection makes use of the following accessories:

- on the vehicle side, a *vehicle inlet* and a *connector*
- on the charging station side, a *plug* and a *socket-outlet*

The cable and plug may be permanently attached to the vehicle (case "A", generally found only in very light vehicles), a detachable cable can be used (case "B", the most common for normal and semi-fast charging), or the cable and connector can be permanently attached to the supply equipment (case "C", typically used for fast charging where heavy cables are used, but posing a higher risk of copper theft for public use).

5.2 Standard Accessories for Mode 1 and 2

For Mode 1 and Mode 2 charging, standard plugs and socket-outlets can be used. Domestic accessories however are not really suited for the heavy-duty operation of electric vehicle charging, characterized by long time operation at near rated current and frequent operation, including disconnection under rated load. This leads to a shorter lifetime of the accessories and to contact problems which may cause hazardous situations. It is thus recommended to limit the rating of the charging equipment using such plugs to a lower value, up to 10 A, their use being confined to small vehicles such as scooters (for which this current level is largely sufficient), as well as for occasional charging of larger vehicles (the "grandma" solution).

A better alternative for Mode 1 or Mode 2 is to use industrial plugs and sockets as defined by the international standard IEC60309-2 [41]. These plugs (in standard blue colour for 230 V, red for 400 V) are widely used, particularly in Europe, for industrial equipment but also for outdoor uses like camping sites, marinas, etc.

5.3 Dedicated Accessories for a.c. Charging

The use of a physical control pilot conductor necessitates the introduction of specific accessories for electric vehicle use. Such plugs and sockets are described in the international standard IEC62196 "Plugs, socket-outlets, vehicle couplers and vehicle inlets—Conductive charging of electric vehicles". Part 1 of this standard [42] gives general functional requirements; it integrates general requirements from the industrial plug standard IEC60309-1 [43] with the electric vehicle requirements of IEC61851-1 [16].

Physical dimensions for a.c. accessories are treated in part 2, was published in 2011 [44]. It does present standard sheets for several types of connectors, vehicle inlets, plugs and socket-outlets:

- Type 1
  The Type 1 single phase coupler is rated for 250 V and 32 A. This solution is featured in SAE-J1772 [45] and based on a proposal made by the Japanese company Yazaki. It is intended to be used as vehicle connector/inlet only, there is no corresponding plug as US charging stations typically work with a Case "C" connection only (Fig. 6).
- Type 2
  Type 2 is a three-phase plug rated for currents up to 63 A, and has two auxiliary contacts. It is illustrated in Fig. 7 and based on a realisation by the German company Mennekes. The need for three-phase accessories was expressed by European car manufacturers and utilities, recognizing the potential benefits of three phase charging and the availability of three phase supply in most European countries. Type 2 also features a connector/vehicle inlet combination (similar but not intermateable with the plug/socket-outlet). The automobile industry is presently mounting both Type 1 and Type 2 inlets on cars and light trucks, depending of the original market of the vehicle. In Europe, both types can thus be found.

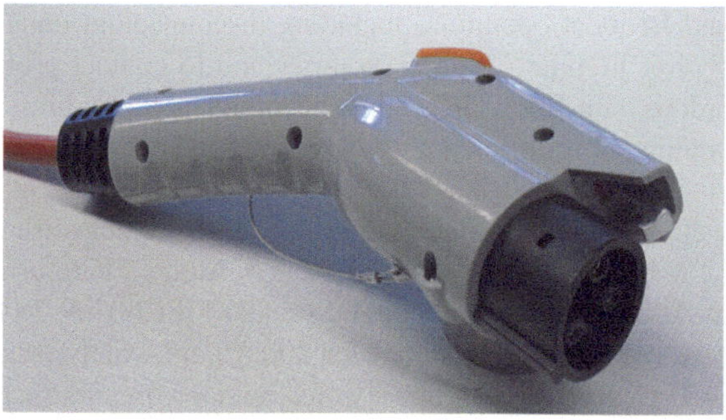

Fig. 6 Type 1 connector

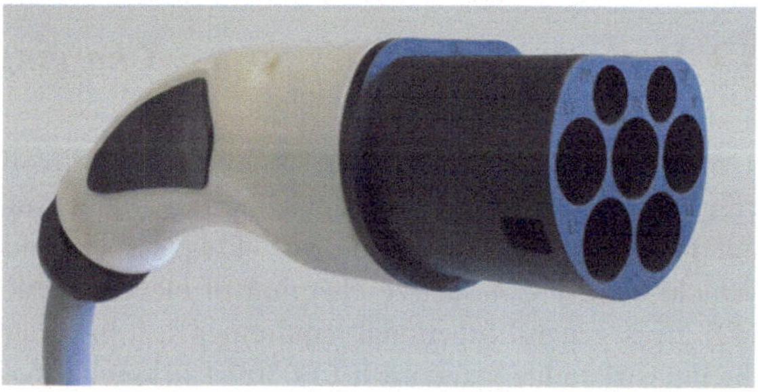

Fig. 7 Type 2 plug

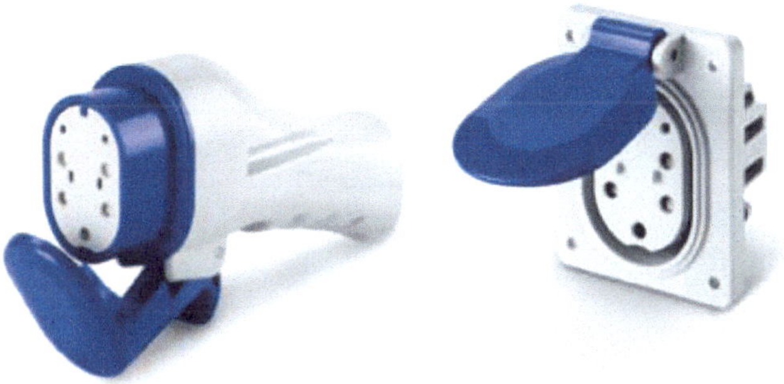

Fig. 8 Type 3 plug and socket-outlet

- Type 3
  Type 3 is also a three-phase type, it is illustrated in Fig. 8 and based on a design by Italian company SCAME further adopted by the "EV Plug Alliance".

The choice of a single type plug (either Type 2 or Type 3) for European charging stations has been a point of discussion. One main difference between Type 2 and Type 3 accessories is the presence of "shutters" on the latter which may be required in some countries by national wiring regulations for socket-outlets in domestic environments. However, shutters are now also available for Type 2.

The proposed European directive on the deployment of alternative fuels infrastructure [46] prescribed the use of Type 2 accessories as the standard solution for Europe. Charging points shall comply with this standard by the end of 2015. Type 3, now still widespread in France and Italy only, is thus likely to be gradually phased out.

5.4 Connectors for d.c. Charging

Connectors and vehicle inlets for d.c. charging are treated in IEC62196-3 which is presently at CDV level [47]. The standard presents three families of connectors: the "CHAdeMO" type of Japanese origin (Fig. 9), the "Combo" type encompassing both a.c. (Type 1 or 2) and d.c. inlets in one unit (Fig. 10), and a Chinese connector type.

It has also been proposed to use Type 2 connectors system with commutable a.c./d.c. pins. Typical use scenarios of the pins are illustrated in Fig. 11. This combined use of a.c. and d.c. on the same pins has however given rise to safety concerns particularly from the electrotechnical industry. The issue is now under consideration in the standards committees.

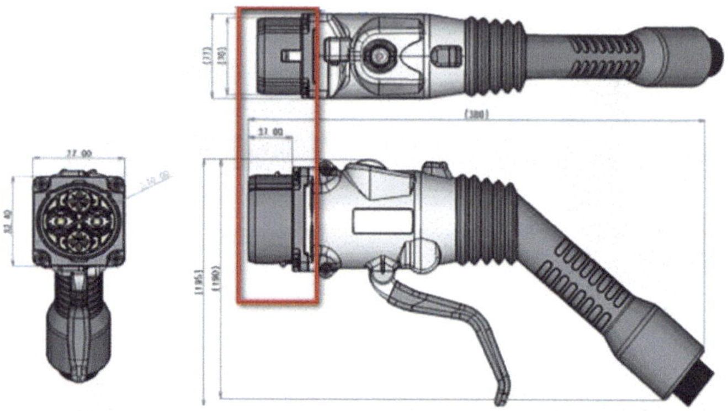

Fig. 9 CHAdeMO connector

Fig. 10 Combo connector example

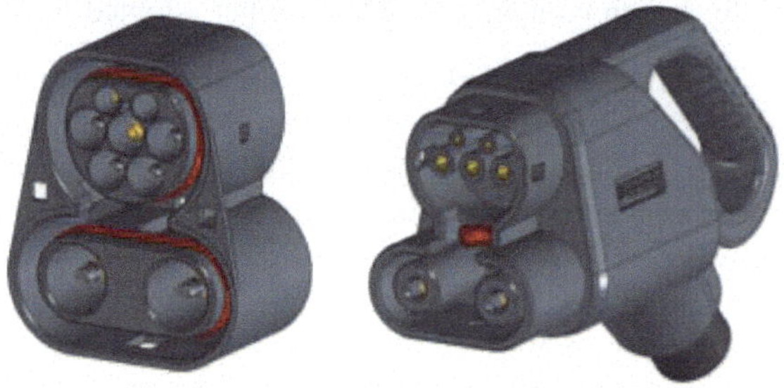

Fig. 11 Combined use of Type 2 connector

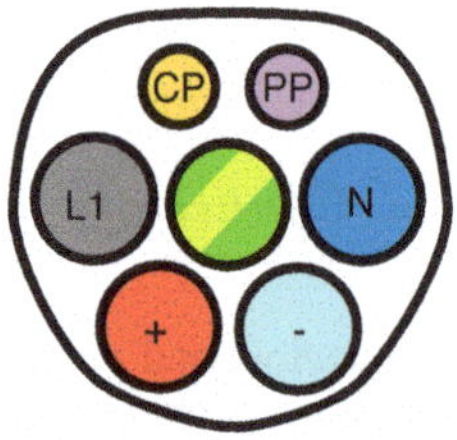

The proposed European directive [46] prescribes the use of "Combo type 2" connectors for d.c. charging stations. Fast charging points shall comply with this standard by the end of 2017.

6 General Vehicle Safety

The general safety of the electric vehicle is covered by the international standard ISO6469, which comes in four parts:

- Part 1 describes the safe installation of the RESS (rechargeable energy storage system) on the vehicle [7]
- Part 2 covers operational safety of the vehicle, focusing on issues which are particular for the electric drive train [48]
- Part 3 considers the electrical safety of the vehicle and the protection of personnel against electric shock [49]
- Part 4, under development, gives requirements for post-crash electrical safety, focusing on the risks present to emergency personnel interventions [8]

7 Conclusions

Electrically propelled vehicles remain a key subject for future standardization work. As with all standardization matters, electric vehicle standards pertain to the three main pillars of the house of standardization (Fig. 12): safety, compatibility and performance.

Safety standards ensure protection against electric shock and other related hazards, as well as controlling electromagnetic compatibility issues, allowing the vehicles to be used safely in all its potential environments.

Compatibility standards obviously refer to the definition of suitable plugs and sockets for electric vehicle charging, but also cover the communication needs of charging and allow the electric vehicle to be deployed in an extended area and the infrastructure to be universally usable.

Performance measurement standards, in the framework of this study, pertain to the measurement of battery performances as well as battery state of charge and state of health.

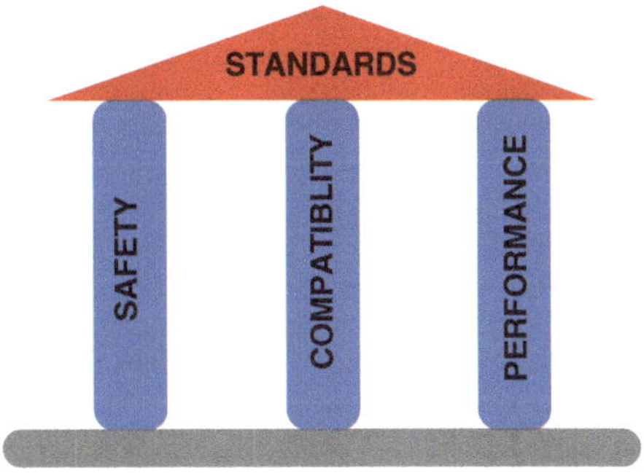

Fig. 12 The house of standardization [50]

A number of issues may be discerned however where a global standard approach has not yet been reached and where further standardization work is needed. Some of these, such as the coexistence of several mutually incompatible connector systems may be resolved relatively easily; there are other subjects however, such as wireless charging technologies or the variety of battery cell designs, which represent a developing technology where a common standard can not yet be set and will be defined by technological maturity. Standardization may however bring an interesting contribution throughout the development stage by defining boundary conditions particularly in matters such as safety.

Intensive work is now being performed by international standardization committees in order to realize unified solutions which will be a key factor in allowing the deployment of electrically propelled vehicles on a global level, highlighting the technical and societal relevance of standardization.

References

1. IEC60254-1 (2005) Lead-acid traction batteries—Part 1: general requirement and methods of test, 4th edn. IEC
2. IEC61982 (2012) Secondary batteries (except lithium) for the propulsion of electric road vehicles—performance and endurance tests, 1st edn. IEC
3. ISO12405-1 (2011) Electrically propelled road vehicles—Test specification for lithium-ion traction battery packs and systems—Part 1: high-power applications, 1st edn. IEC
4. ISO12405-2 (2011) Electrically propelled road vehicles—test specification for lithium-ion traction battery packs and systems—Part 1: high-energy applications, 1st edn. IEC
5. IEC62660-1 (2010) Secondary batteries for the propulsion of electric road vehicles—Part 1: performance testing for lithium-ion cells, 1st edn. IEC
6. CEN-CENELEC (2011) Focus group on european electro-mobility, final report to CEN and CENELEC technical boards in response to Commission Mandate M/468 concerning the charging of electric vehicles, CEN-CENELEC
7. ISO6469-1 (2009) Electrically propelled road vehicles—safety specifications—Part 1: on-board rechargeable energy storage system (RESS). ISO
8. ISO/DIS6469-4 (2014) Electrically propelled road vehicles—safety specifications—Part 4: post crash electrical safety requirements. ISO
9. IEC62485-3 (2010) Safety requirements for secondary batteries and battery installations—Part 3: traction batteries, 1st edn. IEC
10. IEC62660-2, IEC 62660-2 (2010) Secondary batteries for the propulsion of electric road vehicles—Part 2: reliability and abuse testing for lithium-ion cells, 1st edn. IEC
11. ISO12405-3 (2014) Electrically propelled road vehicles—Test specification for Lithium-ion traction battery packs and systems—Part 3: safety performance requirements, 1st edn. ISO
12. IEC62660-3/CDV (2013) Secondary lithium-ion cells for the propulsion of electric road vehicles—Part 3: safety requirements of cells and modules, 1st edn. IEC
13. IEC60254-2 (2008) Lead-acid traction batteries—Part 2: dimensions of cells and terminals and marking of polarity on cells, 4th edn. IEC
14. ISO/IEC-PAS16898 (2012) Electrically propelled road vehicles—dimensions and designation of secondary lithium-ion cells, 1st edn. ISO/IEC
15. IEC61851-1 (2001) Electric vehicle conductive charging system—Part 1: general requirements, 1st edn. IEC TC69 WG4

16. IEC61851-1 (2010) Electric vehicle conductive charging system—Part 1: general requirements, 2nd edn. IEC
17. IEC61851-1/CD (2012) Electric vehicle conductive charging system—Part 1: general requirements, 3rd edn. No. 69/219/CD, IEC TC69 WG4
18. IEC61851-1/CD (2013) Electric vehicle conductive charging system—Part 1: general requirements, 3rd edn. No. 69/250/CD. IEC TC69 WG4
19. ISO17409/CDV (2013) Electrically propelled road vehicles—connection to an external electric power supply—safety specifications, No. 69/263/CD. ISO
20. IEC61851-21-1/CD (2012) Electric vehicle conductive charging systems—Part 21-1: electric vehicle onboard charger EMC requirements for conductive connection to an a.c./d.c. supply, 1st edn. No. 69/222/CD. IEC TC69
21. IEC61851-21-2/CD (2012) Electric vehicle conductive charging system—Part 21-2: EMC requirements for OFF board electric vehicle charging systems, 1st edn. No. 69/220/CD. IEC TC69
22. IEC61851-23 (2014) Electric vehicle conductive charging system—Part 23: d.c. electric vehicle charging station, 1st edn. IEC TC69
23. IEC61851-24 (2014) Electric vehicle conductive charging system—digital/data communication of d.c. charging control between off-board d.c. charger and electric vehicle, 1st edn. IEC TC69
24. IEC61851-3-1,3-2,3-3,3-4 (2012) Electric vehicles conductive power supply system—Part 3.1: general requirements for light electric vehicles (LEV) AC and DC conductive power supply systems—Part 3.2: requirements for light electric vehicles (LEV) DC off-board conductive power supply systems,—Part 3.3: requirements for light electric vehicles (LEV) battery swap systems—Part 3.4: requirements for light electric vehicles (LEV) communication, 1st edn. No. 69/221/NP,69/237/RVN. IEC TC69
25. IEC/TS62763 (2013) Pilot function through a control pilot circuit using PWM modulation and a control pilot wire, 1st edn. IEC TC69 WG4
26. IEC61639-7/FDIS (2013) Low-voltage switchgear and controlgear assemblies—Part 7: assemblies for specific applications such as marinas, camping sites, market squares, electric vehicles charging stations, 1st edn. No. 17D/478/FDIS. IEC SC17D
27. IEC61439-1 (2011) Low-voltage switchgear and controlgear assemblies—Part 1: general rules, 2nd edn. IECSC17D
28. IEC60364-7-722-CDV (2012) Low-voltage electrical installations—Part 7-722: Requirements for special installations or locations—supply of electric vehicle, 1st edn. No. 64/1846/CDV. IEC TC64
29. IEC61980-1/CD (2012) Electric vehicle wireless power transfer systems (WPT)—Part 1: general requirements, No. 69/236/CD. IEC TC69
30. IEC62840/NP (2012) Electric vehicle battery exchange infrastructure safety requirements, No. 69/217/NP. IEC TC69
31. Directive 2004/108/ec of the european parliament and of the council on the approximation of the laws of the member states relating to electromagnetic compatibility and repealing directive 89/336/eec, EU OJ L 390, 31 Dec 2004
32. Directive 2004/104/ec of 14 October 2004 adapting to technical progress council directive 72/245/eec relating to the radio interference (electromagnetic compatibility) of vehicles and amending directive 70/156/eec on the approximation of the laws of the member states relating to the type-approval of motor vehicles and their trailers, EU OJ L337, 13 Nov 2004
33. ISO/IEC15118-1/PRF (2013) Road vehicles—Vehicle to grid communication interface—Part 1: general information and use-case definition, 1st edn. JWG ISO TC22 SC21 / IEC TC69
34. ISO/IEC15118-2/DIS (2012) Road vehicles—vehicle-to-grid communication interface—Part 2: technical protocol description and open systems interconnections (OSI) layer requirements, 1st edn. JWG ISO TC22 SC21/IEC TC69

35. ISO/IEC15118-3/DIS (2013) Road Vehicles—Vehicle to grid communication interface Part 3: physical layer and data link layer requirements, JWG ISO TC22 SC21/IEC TC69/IEC, 2013
36. ISO/IEC15118-4/NP (2012) Road Vehicles—vehicle to grid communication interface Part 4: Network and application protocol conformance test. JWG ISO TC22 SC21/IEC TC69
37. ISO/IEC15118-5/NP (2012) Road Vehicles—vehicle to grid communication interface Part 5: Physical layer and data link layer conformance test. JWG ISO TC22 SC21/IEC TC69
38. ISO/IEC15118-6/AWI (2013) Road Vehicles—vehicle to grid communication interface—Part 6: general information and use-case definition for wireless communication. JWG ISO TC22 SC21/IEC TC69
39. ISO/IEC15118-7/AWI (2013) Road Vehicles—vehicle to grid communication interface—Part 7: network and application protocol requirements for wireless communication, JWG ISO TC22 SC21 / IEC TC69
40. ISO/IEC15118-8/AWI (2013) Road Vehicles—vehicle to grid communication interface—Part 8: physical layer and data link layer requirements for wireless communication, JWG ISO TC22 SC21/IEC TC69
41. IEC60309-2 (2012) Plugs, socket-outlets and plugs for industrial purposes—Part 2: dimensional interchangeability requirements for pin and contact-tube accessories, 4th edn. IEC
42. IEC62196-1 (2010) Plugs, socket-outlet and vehicle couplers—conductive charging of electric vehicles—Part 1: charging of electric vehicles up to 250 A a.c. and 400 A d.c., 2nd edn. IEC
43. IEC60309-1 (2012) Plugs, socket-outlets and plugs for industrial purposes—Part 1: General requirements, 4th edn. IEC
44. IEC62196-2 (2011) Plugs, socket-outlet and vehicle couplers—conductive charging of electric vehicles—Part 2: Dimensional interchangeability requirements for pin and contact-tube accessories with rated operating voltage up to 250 V a.c. single phase and rated current up to 32A, 1st edn. IEC SC23H
45. SAE J1772 (2012) Electric vehicle and plug in hybrid electric vehicle conductive charge coupler. SAE
46. European Commission (2013) Proposal for a directive of the European Parliament and of the Council on the deployment of alternative fuels infrastructure, No. 2013/0012(COD)
47. IEC62196-3/CDV (2012) Plugs, socket-outlet and vehicle couplers—conductive charging of electric vehicles—Part 3: dimensional interchangeability requirements for d.c. and a.c./d.c. pin and tube-type vehicle couplers, 1st edn. No. 23H/292/CDV. IEC SC23H
48. ISO6469-2 (2009) Electrically propelled road vehicles—safety specifications—Part 2: vehicle operational safety means and protection against failures. ISO
49. ISO6469-3 (2011) Electrically propelled road vehicles—Safety specifications—Part 3: protection of persons against electric shock. ISO
50. Van den Bossche P (2010) Matching accessories: standardization developments in electric vehicle infrastructure. In: EVS-25

Barriers and Opportunities for SMEs in EV Technologies: From Research to Innovations

Neil Adams, Christopher Pickering, Richard Brooks and David Morris

Abstract This report has been produced as part of the FP7-funded research project INTRASME (Innovative Transport SME Support Action) which aims to improve the capacity and capability of European SMEs to develop and implement products more rapidly in the low carbon transportation sectors. This report presents results of work on how SMEs acquire new technologies and develop new products and services and the value SMEs get from participating in EU R&D Transport programmes, focusing on the barriers SMEs face in exploiting their innovations and how these can be overcome. The report identifies strategies associated with the successful commercialisation of technology, and produces recommendations for the European Commission on support for SMEs.

Keywords Electric vehicles · Smart mobility · Low carbon transport · SMEs · Barriers · Commercialization · Productionisation

1 Introduction

According to the European Roadmap for Electrification of Road Transport [1], up to 5 million electric cars could be in use by 2020 in Europe, rising to 15 million electric cars in 2025. If one includes e-bikes, e-scooters and other e-mobility systems the

N. Adams (✉) · C. Pickering
Innovation Bridge Consulting Ltd, Wyche Innovation Centre Walwyn Road, Upper Colwall, Malvern, Worcs WR13 6PL, UK
e-mail: nadams@innovationbridge.co.uk

C. Pickering
e-mail: cpickering@innovationbridge.co.uk

R. Brooks · D. Morris
Coventry University Enterprises, Priory Street, Coventry CV1 5FB, UK
e-mail: ab3689@coventry.ac.uk

D. Morris
e-mail: d.morris@coventry.ac.uk

B. Müller and G. Meyer (eds.), *Electric Vehicle Systems Architecture and Standardization Needs*, Lecture Notes in Mobility,
DOI 10.1007/978-3-319-13656-1_2

figure will be far higher, with up to 30–40 million electrically powered mobility vehicles in use in 2020 [2]. These new systems offer new opportunities for SMEs as described in the ICT for the Fully Electric Vehicle Roadmap [3] based on changes to the supply chain and emerging business models. While the internal combustion engine (ICE) is an extremely complex system whose control depends on components and software packages in the hands of a few large organisations, the management of one or more electrical motors is much less demanding and is accessible to many new organisations including SMEs. Similarly the management of the production of battery packs may not be a monopoly of a few large companies. The investments for a full production plant of light and heavy quadricycles can be as much as 15–25 times lower than the investment needed for a small conventional M1 ICE city car. The introduction of these new forms of mobility could be driven by new players acting much faster than large OEMs are used to. As the 2011 EU Transport White Paper states "it is clear that SMEs will have a pivotal role to play in this sector, being quick to adapt to new and emerging technologies in the sector".

There appears to be significant scope for SME innovation in this domain, but significant barriers remain. According to JRC analysis, innovators in the transport sector bear the risk that their up-front investments will not deliver a satisfactory return for reasons that include [4]:

- High capital intensiveness, reinforced by problems of financing;
- Uncertainty in market demand (which limits the incentive to innovate);
- Complex innovation systems that require coordinated innovation efforts and innovation speeds between all players (e.g. vehicle/fuel/infrastructure/consumer), including industry, academia and governments;
- Markets that are dominated by established enterprises and therefore make it difficult for newcomers to enter.

The transition of new technologies from research through to deployment in the market is a risky step that often fails: this transition is often called the 'valley of death'. This paper looks at how SMEs acquire new technologies and how EU funded R&D projects could help SMEs in making this transition.

Although there appears, in principle, to be a good opportunity for SMEs in this domain, which EU funded R&D programmes could help them to address, it is important to look at the SME experience from their point of view to understand the relevance and benefits of such programmes to them and how they could be improved. Large organisations who work with SMEs in EU projects also develop a good understanding of the barriers they face and how such projects can help to address them, and their insights are also captured in this paper.

This paper has been produced as part of the FP7-funded project INTRASME (Innovative Transport SME Support Action) which aims to improve the capacity and capability of European SMEs to develop and implement products more rapidly in the low carbon transportation sectors.

2 Methodology

The focus of the INTRASME project is on electro-mobility (matching most of the activities in the European Green Vehicles Initiative apart from alternative fuels [3]) and smart/intelligent mobility. The topics addressed are:

- Low Carbon (Land) Vehicles (Hybrid and Full Electric)—Electric Vehicle (EV) Technologies and Fuel Cells for range extension.
- Light Aircraft—Electric Aircraft.
- Electric water vehicles.
- Smart Mobility: End to End Journey Management—Focus on systems needed to support electromobility, e.g. Intelligent Transport Systems, Smart Grid, rather than all co-modality projects which may have more limited economic impact for companies.
- Enabling Technologies that can be applied to the above low carbon transport/ electromobility applications, e.g. Supercapacitors.

The INTRASME project has analysed how firms acquire and develop low carbon transport technology in different regions of Europe, by identifying the processes involved in developing a new technology/product application through to the early stages of commercialisation (i.e. pre-production/small volume sales), as follows:

- Data on SME technology development processes, practices and experiences has been gathered through an interview process of 57 SMEs from four target regions listed below and also from other EU countries to ensure that results from the survey can be generalised and applied to Europe as a whole:
  - UK, West Midlands
  - Poland, Warsaw
  - Italy, Piedmont
  - Bulgaria, Ruse
- These SMEs on occasion have participated in EU R&D programmes but more often have not participated in EU or national programmes or have participated in national programmes only. Comparing their views with SMEs that are involved in EU R&D projects has provided context for understanding the level and effectiveness of SME participation in EU R&D programmes.

Since INTRASME has been set up to help innovative SMEs enter supply chains, the following types of EU R&D projects were studied to identify SMEs involved in EU R&D projects to interview:

- Projects with high innovation potential with significant involvement of SMEs.
- Projects that are still on-going or recently completed where exploitation and dissemination support could still be helpful.
- Projects representative of different types of R&D Projects from different EC programmes, and of the different challenges faced by SMEs.

A short list of projects based on the above criteria was produced by searching through EC project data sources (primarily CORDIS and project websites). The projects were analysed to identify those with significant innovative SME involvement, either as an important partner supplying technology or in a few cases acting as project coordinator. The list was also filtered to ensure coverage of the low carbon transport technology categories identified above. 38 SMEs were identified and targeted for interviews, including 4 SMEs acting as coordinators of EU projects. The SMEs short-listed for interviews were contacted and a good response rate of about 55 % was achieved with good coverage of EU countries. These SMEs are/have been involved in 31 EC R&D projects.

Coordinators and other participants also involved in the projects and in exploiting the results, e.g. Original Equipment Manufacturers (OEMs), Tier 1s, Research Institutes were also selected for interview to capture their views of SME involvement. 25 Coordinators/other organisations that were not SMEs but were involved in EU projects with SMEs were identified and targeted to approach for interviews, and 14 organisations involved in 21 EC R&D projects were interviewed.

Data on SME decisions, actions and experiences (and those of organisations working with SMEs in R&D projects) were captured during the course of the project in structured questionnaires. These data were logged and aggregated. By providing a structure that provides a degree of consistency of results over a larger sample it is possible to construct generalised observations about common experiences.

3 SME Acquisition and Development of New Technologies and Impact of EU R&D Projects

3.1 Perception of Market Opportunities

Participants were asked why they saw an opportunity to enter the EV and Smart Mobility markets, in order to assess why they sought to acquire a particular technology. Most commonly cited reasons are listed below:

- Personal interest in the technology or application of the technology
- The intrinsic benefits of the technology (ecology, performance, safety)
- Opportunity to open a new market
- Future market growth trends
- An environmental philosophy

Prominent amongst the answers was a sense that the EV sector presents an opportunity to open a new market where there is opportunity for future growth.

Participants were also asked why they believed EV products were being adopted or not by end users. See Fig. 1.

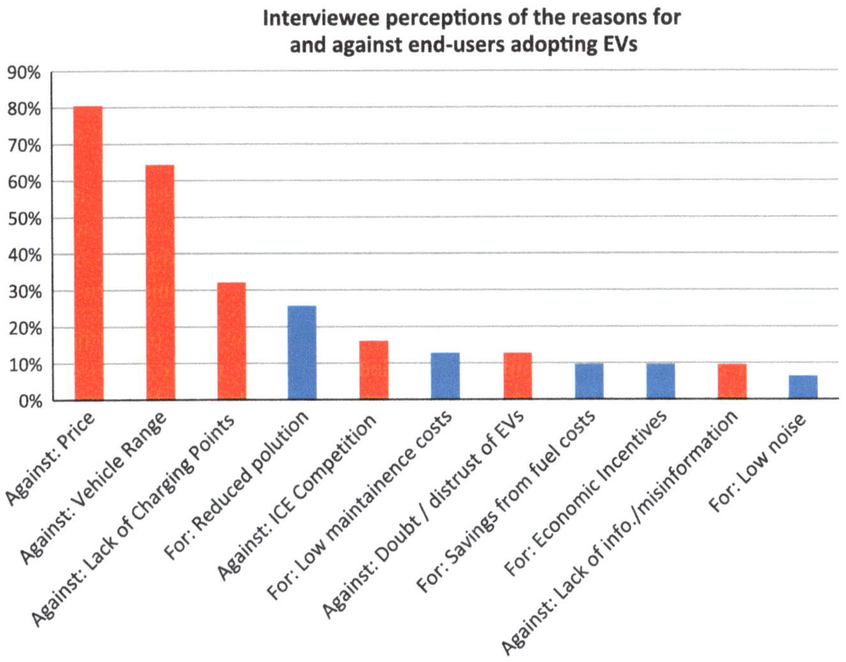

Fig. 1 SME views of reasons for and against end-users adopting EVs

Negative reasons greatly outweighed positive reasons. Amongst the reasons for non-adoption price, battery range and infrastructure featured highly, followed by competition from conventional vehicles and uncertainty on behalf of the consumers. Positive reasons included sustainability, reduced maintenance costs, savings from reduced fuel costs and economic incentives. These views reveal a perception that the technology and business case for EVs is still under-developed. This presents an opportunity for new entrants to benefit from technological improvements but also underlines the risk that unless these issues are resolved there may be insufficient demand to amortise costs.

Several of the reasons for and against adoption also point to the role of external regulations (policy drivers to reward use of less polluting vehicles), intervention (economic incentives and disincentives) or the influence of other actors (the availability of charging points, competition from ICE vehicle producers).

In summary, EV businesses perceived considerable scope for technological change to contribute to and benefit from growth in a new market. The inverse, however, is that unless technological challenges (as relates to price, weight and performance) and issues external to the market (such as the availability of charging points) are resolved, there are doubts in the supply chain about the viability of the value proposition to the consumer.

3.2 SME Acquisition of Technology

Figure 2 provides an overview of how SMEs source ideas

The most important sources of ideas, in order of responses, are from management and in-house R&D, followed by partners and regional clusters, market trends and commercial partners. This demonstrates the central role of entrepreneurs (management and founders) as the drivers and link between networks of practical know-how (partners and regional clusters) and networks of market knowledge (commercial partners). This balance of supply and demand (factor) shifts as companies grow in size.

- Micro businesses are most likely to draw on the know-how of partners and regional clusters, but are unlikely to be led by: market trends (4 %) or sales (0 %). Medium-sized companies by contrast are more likely to develop ideas in response to: market trends (57 %) and sales (14 %), and companies that have successfully commercialised a product are significantly more influenced by commercial customers (ranging from 16 to 47 %). This suggests that as organisations grow, demand factors play a greater role in creation of new ideas.
- Collaborative R&D features more highly as a source of new ideas for medium sized businesses (29 %) than for micro-businesses (0 %). This is likely to be associated with the greater access to EU R&D funds: where 33 % of medium-sized businesses have participated in projects against 12 % of micro-businesses.
- Sales and end-users do not feature as a major source of ideas for any size of company.

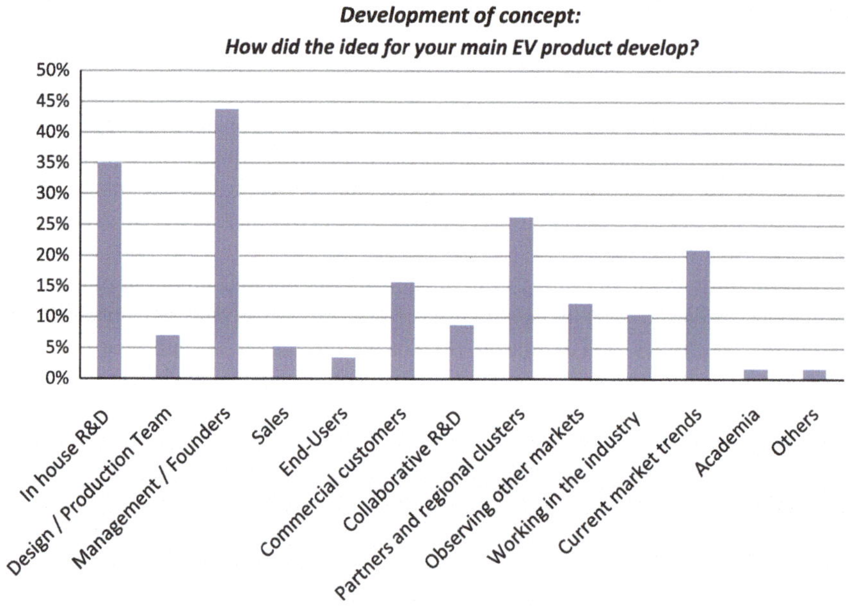

Fig. 2 Development of the concept

3.3 Impact of SME Involvement in EU R&D Projects

The SMEs involved in EU R&D projects, covering the full range of sizes from micro through to medium, gave their views on the effectiveness of collaboration in 26 EC collaborative projects in which they are involved, and the results are shown in Table 1.

In half of the projects the SMEs felt that the collaboration was highly effective while in just over a quarter of cases the collaboration was viewed as moderately effective, for reasons discussed below. The effectiveness of collaboration was only rated as low in one case (4 % of projects).

Concerns raised by several SMEs included:

- Fewer participants would be more effective—the more participants, the more project time goes to information gathering and communication.
- Cases of OEMs doing in-house solution development and less integrated into projects than other participants.
- Slow OEM decision-making.

Once SMEs are in EC projects they are generally happy with the level of collaboration. They need help to get into projects if they are not already familiar with them, and once in a project they value the help provided by coordinators who support them in carrying out the required administration tasks. Collaboration increases in effectiveness as partners get to know each other. Benefits provided to SMEs that they highlighted in the interviews included:

- Higher profile and increased credibility from the moment a project is awarded and publicised (in one case an SME was saved from bankruptcy by orders that followed the publicising of an EC project award).
- Acquiring new knowledge and skills.
- Broader market view.
- Strengthening collaboration with large industries and research institutes.
- FP7 works well as the researchers/engineers interact at the working level, which is a key benefit, enabling SMEs with new concepts to learn what OEMs look for in terms of production-ready technology, e.g. need to carry out formalised failure analysis, production tolerances, etc.
- Possibility of developing future products/services and working with project partners who can help them.

Table 1 SME views of the effectiveness of collaboration in EC R&D projects

Effectiveness of collaboration—SME view	Highly effective	Moderately effective	Low effectiveness	Too early to judge
Number of projects	13	7	1	5
Percentage of projects (%)	50	27	4	19

- Provides funding allowing SMEs to risk employing people to develop new solutions.
- Supporting SME participation in research projects through a higher funding rate than for large companies (as it is at present) helps a lot.

The interviews with SMEs were also used to establish common problems and barriers SMEs face and to give them the opportunity, based on their experience in the EU projects, to suggest improvements. In addition, interviews with non-SME Coordinators or other key participants were carried out including similar questions to be answered from their perspective of the SME involvement in their projects. These other participants generally had wide experience of EU projects and working with SMEs. The main barriers identified by SMEs involved in EC R&D projects to developing and exploiting their innovations, validated by coordinators and other non-SME organisations in such projects, are listed below in Table 2, together with the frequency with which they were raised by SMEs.

The impact assessment report that looked at SME participation in Framework 5 and 6 in March 2010 [5] found that SMEs, while reporting positive impacts from participation in the project, were generally not optimistic about exploitation. Similarly for the SMEs in this study the issue of linking to larger partners and getting effective exploitation of their ideas is their dominant concern.

Finance and making the business case for investment is another major concern of innovative SMEs participating in EC R&D projects. Accessing finance is particularly difficult at present, and with uncertain demand from customers, is exacerbated by existing electric vehicles being seen as too expensive for general adoption and hybrid technology as already available and adequate to meet the demand. Hence investors often see no commercial reason to support new EV technologies.

SMEs are felt to be a good source of innovative ideas by other project partners, but SMEs felt that they could benefit much more from links to Universities/Research Institutes. SMEs sometimes feel disconnected from research in academia, recognise the need to develop collaborative relationships, but often lack the time to do so.

Table 2 Main barriers cited by SMEs to developing and exploiting their innovations

Barriers in order of importance	Number of SMEs	Proportion of SMEs (%)
Linking to exploitation partners—OEMs, Tier 1s/Tier 2s ETC	12	57
Finance and business case (including market need)	10	48
Innovation and links to universities/research institutes	8	38
Productionisation	6	29
Need to join clusters for critical mass	6	29
Need to understand EC programmes and bid processes and for guidance and support	4	19

Taking working prototypes through to production is seen as a major challenge: bringing new technologies into production is too expensive for many SMEs.

SMEs value being part of a larger community or network, especially micro-SMEs, which gives them access to tools and knowledge they need, and sometimes need access to such networks to understand and exploit new low carbon transport opportunities. There are interesting examples such as the Torino e-District in Italy where SMEs are joining together (and with manufacturing partners) to aggregate capability to achieve critical mass and jointly pursue new opportunities in electric vehicles.

Even the SMEs interviewed in this study, who are already involved in EC R&D projects, still felt they lacked understanding of EC programmes and how to get into new projects of value to them and their customers. The biggest barrier to SME involvement on EC R&D projects is the perception of the difficulties and problems both in bidding and participating in such projects, some but not all of which are justified. Once SMEs are in EC projects they are generally happy with the level of collaboration, and value the help provided by coordinators who support them in carrying out the required administration tasks.

3.4 Strategies Deployed by Small Companies that Have Successfully Overcome Barriers to Commercialise Their Products

68 % of the SME firms sampled stated they had commercialised their product(s) and, of this number, one third supplied outside their home region/nation. In all cases companies identified product performance (which often included innovative features) and price as essential elements in commercialisation. However a number of other key factors were identified:

- The importance of establishing an early lead in a niche market. Examples include: material handling vehicles, mobility vehicles, EV scooters, EV conversion kits etc.
- A strong emphasis on flexibility, customisation and responding to customer needs. Successful firms were often able to offer a wide range of services or adaption either through a diversity of internal competencies or leveraging the benefits of local supply clusters. The importance of responding quickly to customer needs was seen as essential in winning contracts; in one case the principle product was entirely redesigned and applied to an entirely separate sector.
- Exploiting comparative advantages of low labour costs and the competitive advantages of being the first in the field. Both of these factors were considered beneficial in Bulgaria and Poland, and a competitive advantage was identified in Italy, the UK and Finland where the absence of local competition in a sector enables the company to establish a dominant position.

- An open approach with large customers and an emphasis on protecting IP through superior know-how and delivery is important.

For firms that are at a pre-commercialisation or low volume production stage the key challenges are:

- Financing the R&D phase and the production and commercialisation phase: the first relates to funding R&D and prototyping stages, where firms are usually reliant on their own capital; the second relates to marketing low volume/high unit cost capital products to potential suppliers, where the costs of development need to be offset before investments can be made in mass production.
- Managing through uncertainty and promoting a convincing value proposition where all stages of the supply chain are unsure about the degree of market demand for EV products and likelihood that demand will offset the development costs of improving technology.
- Expanding beyond a local region where supply and demand factors are favourable.

These observations have two significant implications for understanding the product development process in the electric vehicle market:

- The product development process most frequently described by SMEs is strongly orientated around the technical development of the product. In firms where a product has been successfully commercialised, a focus on technical performance is combined with the ability to quickly customise these competencies to the needs of the market. This is reflected by the need expressed by companies for more marketing/brokerage support.
- An understanding of a firm's position within the surrounding industrial structure is essential. Almost all of the companies that had commercialised a product supplied B2B (business to business) and half of the original product ideas came from commercial partners. Clusters of similar companies can enable SMEs to share know-how and market knowledge as well as draw on a diverse range of competencies and quickly customise an offer. The presence of sympathetic regional support can help offset development and promote confidence in the region about the technology's future. The presence of a developed and localised supply chain enables SMEs to diversify their product range and reduce risk (examples include supplying to the commercial and higher education sectors or to the EV and ICE sectors) where SMEs are able to establish continuing relationships.

The manufacturing of Micro EVs does not necessarily require the expertise of the classical OEMs and in Europe there are groups of organisations (clusters), some represented by INTRASME partners, e.g. Torino e-District, that are developing the capability to develop supply chains capable of quickly responding to market demands. National and regional Government policy-maker support is required for such clusters to launch initiatives in local production or parts assembly, but Governments and funding agencies need help to recognise the potential for Micro

EVs that can be produced with much lower investments than are necessary for conventional vehicles.

- The INTRASME project has already stimulated initial work to develop formal Low Carbon Transport clusters in the Midlands (UK) and Ruse (Bulgaria) and also work to broaden existing initiatives around Warsaw (Poland) and Torino (Italy).
- The INTRASME project has also started the formation of an EU Super-Cluster, acting to promote the role of SMEs and regional clusters in the emerging market of Electromobility, at the Smart Mobility World conference in Torino in September 2013.

The overall SME view of the current EV/ITS market is as follows:

- The market is immature with some well-developed niches of activity e.g. Electric Golf Buggies, but low market penetration in areas such as automotive and aircraft
- Collaborative R&D, e.g. with Universities, is important in early product development stages
- Two 'valleys of death' exist in the development of products:
  - Feasibility stage: funding is usually accessible for early stage R&D (from own funds, public support).
  - Production stage: funding gap and a greater challenge for SMEs.

Against this background, SMEs that successfully commercialise their products use the following strategies:

- Acquire a strategic partner to exploit or invest in the technology.
- Offer to serve different markets with their product/competency.
- Develop services to cross-subsidise their main product.
- Focus on highly customised products/systems for customers.

4 Recommendations to the European Commission

The INTRASME project has made a number of recommendations to the European Commission that will assist SMEs in taking their innovations to market. Space only permits a brief description of some of the recommendations but more details are available in the INTRASME reports (www.intrasme.eu). Key recommendations include:

(1) **Mid-way Review of Exploitation Plans.**
Make review of exploitation plans a more important part of each project, carrying out a strict review of exploitation plans halfway through EC projects that are aimed at taking new technologies to market.

(2) **Productionisation Projects.**
The EC should consider supporting projects to show that novel ideas are production ready, and help manufacture successful prototypes. This would not require more EC R&D investment, but a re-balancing of investment towards larger productionisation projects at higher technology readiness levels away from smaller Specific Targeted Research Projects (STREPs) within the same overall budget. These projects will need to be a different type of project to STREPs and may involve different types of EC partners, e.g. banks, business angels, business development companies and could even offer the opportunity for the EC to play a role, e.g. to take equity. Any support should be conditional on future manufacturing in the EU, which will also benefit EU SMEs.

(3) **Shape EU EV policy around regional policy.**
EU EV policy must be shaped around regional policy and coordinated with national governments. The production of electric passenger vehicles is by necessity the culmination of inputs from a wide range of supply chain actors and other interested actors. For emerging EV producers the vitality of local clusters is important for sharing knowledge and resources and finding launch customers. Large OEMs, Tier 1 suppliers, or large fleet operators may act as primary customers for a range of EV suppliers, but are likely to be influenced only by policy directions made at an EU or national level.

5 Conclusions

This paper describes how SMEs acquire technology and develop new products and services across the EU, and the effectiveness of EU R&D projects in helping SMEs exploit their innovations. Key points are:

- Companies view the current EV market as immature with some well-developed niches of activity, but with a low level of market penetration in areas such as aircraft and the automotive market.
- Interest in entering the EV market is motivated by factors including personal interest in the technology and the perception of the market as at an early stage development, where there are opportunities for new entrants.
- Collaborative R&D, the exchange of knowledge between partners and the involvement of universities is important in the early product development stages.
- SMEs involved in EU R&D Projects and partners working with these SMEs identified barriers to exploitation of SME innovations, of which the most significant was the difficulty of linking to exploitation partners, and the INTRASME project has made recommendations to the EC on how they can be addressed.
- There are two 'valley of death' stages in the development of EV products: at the development stage and at the production stage. SMEs typically find the second stage to be the greater challenge as funding exists for early stage R&D but

finding a route to market is more challenging where an existing supply chain may not exist.

- Firms that have been successful in commercialising products employ the following strategies:
  - acquire a strategic partner to exploit or invest in the technology;
  - offer to serve different markets with their product/competency;
  - develop services to cross-subsidise their main product;
  - focus on highly customised products/systems for customers.

Overall the development of a lead market for EV production requires the coordination of a number of regional actors in the supply chain system. A focus on collaborative R&D is extremely important, but without a defined customer and structured set of subsidies or incentives, R&D support may benefit other industries or regions, or simply not reach production. It is by the identification of a route to market (often through the support of a number of existing key companies) that this 'valley of death' stage can be overcome.

To maximise their impact, further support for the exploitation of SME innovations is required from the European Commission (as well as national agencies). This paper makes a number of recommendations to the European Commission in relation to the exploitation of EU R&D Project results, EU R&D project mechanisms, and policy. More details of the work of the INTRASME project are available at http://www.intrasme.eu/.

Acknowledgments The authors would like to thank Pietro Perlo of IFEVS, Torino for his insightful contributions. The results presented in this paper have received funding from the European Union's Seventh Framework Programme (FP7/2007–2013) under grant agreement no 314587, project name: INTRASME.

References

1. European roadmap: electrification of road transport—EPRTRAC, EPOSS, SMARTGRIDS—June 2012
2. ICT4FEV D3.2—roadmap ICT for the Fully Electric Vehicle—30 Oct 2012
3. Multiannual roadmap for the contractual PPP "European Green Vehicles Initiative" draft version 1.0 (26 July 2012) for stakeholders consultation
4. Mapping innovation in the European transport sector—JRC—EUR 24771 EN 2011
5. SMEpact—impact assessment of the participation of SMEs in the thematic programmes of the fifth and sixth framework programmes for RTD—final report, March 2010

Part II
Scientific Papers

OpEneR—Approaching an Optimal Energy Management for Fully Electric Vehicles

Kosmas Knödler and Sylvain Laversanne

Abstract Today's electric vehicles are facing the driver's anxiety for reachability of a chosen destination. This is caused by a rather limited driving range resulting from a still open and sophisticated challenge, i.e. the development of batteries with sufficient high power and energy content, and with high lifetime expectancy. The European publicly funded project OpEneR was established to follow an alternative but somehow complementary way towards a better situation. Driving strategies and assistance systems are being developed to increase efficiency, driving range, and safety. All functions are handled by a novel overall energy management system merging data from multiple on- and off-board data sources. This article gives an outline of OpEneR and its working fields while the project is still ongoing.

Keywords Electric vehicle · 4WD · Efficiency · Safety · ADAS · MiL · SiL · Co-simulation · Car-to-Infrastructure · Car-to-Car · VDC · CRBS · ESP®*hev* · TCS · GPS · Radar · Video · Intelligent transportation system

1 Introduction of the Project

From Bosch's original idea of setting up OpEneR (Optimal Energy consumption and Recovery based on a system network, [1]) until the official project start in May 2011, less than 1 year passed. In the meanwhile contacts were made between Bosch and PSA first. After their agreed mutual interest in a cooperation both partners

K. Knödler (✉)
Robert Bosch GmbH, Chassis Systems Control, Engineering Advanced Development (CC/ENA), Postfach 1355, 74003 Heilbronn, Germany
e-mail: Kosmas.Knoedler@de.bosch.com

S. Laversanne
PSA Peugeot Citroën Automobiles SA DRIA, Route de Gisy, 78140 Vélizy-Villacoublay, France
e-mail: Sylvain.Laversanne@mpsa.com

B. Müller and G. Meyer (eds.), *Electric Vehicle Systems Architecture and Standardization Needs*, Lecture Notes in Mobility,
DOI 10.1007/978-3-319-13656-1_3

Table 1 OpEneR partners and their expertise contributions to the project

Partner	Expertise
Robert Bosch GmbH (coordinator), Germany	Regenerative vacuum free braking system, surround sensing systems (radar, video), electric motors, power electronics, active acceleration pedal, E/E-architecture
Peugeot Citroën Automobiles SA, DRIA, France	EV specification and build-up with subsystems like batteries, energy management integration, vehicle operation strategy
Robert Bosch Car Multimedia GmbH, Germany	Navigation system with 3D maps and eHorizon, eco routing, range management
AVL List GmbH, Austria	Complete simulation environment (SiL, MiL, HiL)
CTAG—Centro Tecnológico de Automoción de Galicia, Spain	New HMI, Car-to-Infrastructure, Car-to-Car communication
FZI—Forschungszentrum Informatik, Germany	Requirements and use cases analysis, operation and driving strategies, simulation

looked for additional institutions with required expertises and own ideas for an optimal energy management system for electric vehicles (EV). Bosch went successfully for AVL (Austria), FZI (Germany), and Robert Bosch Car Multimedia GmbH (Germany). PSA invited CTAG (Spain) into the consortium.

A powerful group of six partners with suitable expertise (see Table 1) was established and intensified the definition of project contents and objectives. From the proposal submission until the end of negotiation only 4 months passed. Finally, the project was defined by the formal characteristics of 3 years runtime (May 2011 until July 2014), € 7.7 M budget and € 4.4 M European contribution.

1.1 Objectives of the Project

OpEneR is developing driving strategies and assistance systems that significantly increase EV efficiency, driving range and safety. This is achieved by merging data from on-board and off-board information sources and by using them in an overall energy management system. A particular focus lies on an optimal cooperation between the electric drivetrain and the regenerative braking system, supported by data from radar, video, satellite navigation, Car-to-Infrastructure (c2i) and Car-to-Car (c2c) systems. The varying subsystems are state-of-the-art and partially new developed components, but most of them do not use fully new developed technologies. Roughly spoken, the project takes “off-the-shelf” subsystems and establishes on that basis a sophisticated energy management system. In this sense OpEneR goes in a somehow complementary direction compared to many other publicly funded research projects.

Figure 1 depicts the project scope: an OpEneR prototype is equipped with the overall energy manager with several new functions that allow the vehicle to drive in the most energy efficient way. To some extent this can be supported by individually

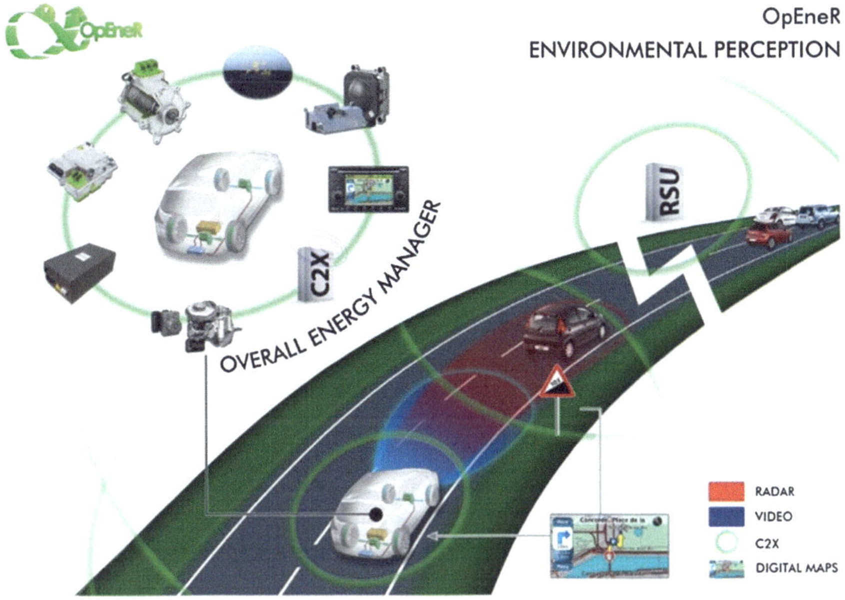

Fig. 1 OpEneR research fields and objectives in a nutshell

running all on-board subsystems in operating points with highest efficiencies. For example the battery system works best when well controlled and conditioned by the Battery Management System (BMS) and by its thermal management. Moreover, it is obvious that there exist many dependencies between the subsystems, also depending on driving scenarios. To achieve a higher degree of overall efficiency it is extremely important to consider the network of several sub-systems and not a single isolated subsystem itself. In this sense high potential for improvement particularly comes from powertrain components (electric motors, power electronics, and clutches), braking system components, battery system (both in propulsive and braking situations) and from HVAC (Heating, Ventilation and Air Conditioning). In addition the driver himself strongly influences the energy consumption of a vehicle. Therefore a well defined way to inform and to instruct the driver via a suitable Human Machine Interface (HMI) is of particular importance, as well as proposing optimally efficient assistance driving systems. Last but not least the vehicle itself can not be considered as driving alone on a selected route. There are many other road users and infrastructure systems, e.g. traffic lights and traffic signs, with mainly limiting impact on the vehicle's degrees of freedom. The environmental perception of the OpEneR prototype is therefore also of high relevance. For near and also visible surrounding objects, systems like radar, ultra sonic and infrared sensors or video cameras can be used. Enhanced radar based Adaptive Cruise Control (ACC) can guarantee the energy efficient approach towards a slower car in front (see black car in Fig. 1). Furthermore c2c communication units can help. When looking in further distances and also on upcoming traffic and environmental situation there is a need for other information sources providing a foresight. A suitable on board system is the satellite navigation system that today considers traffic information (traffic jam in Fig. 1) and that more and more uses 3D map data (slope sign in Fig. 1). Other and partially redundant information can be gathered via c2i (Road Side Units (RSU) in Fig. 1), and again c2c communication lines. RSU can broadcast information on traffic situation, construction places, variable speed limits and weather conditions. In near future the penetration of such systems most likely will quickly increase.

1.2 OpEneR Technology Levels

When defining the project the consortium already decided to go for real demonstration of the OpEneR prototypes with newly developed systems and functions on public roads. As interim steps of course, intensive tests on partners' proving grounds are necessary in order to guarantee a safe vehicle handling and to achieve the required public road release by relevant partners PSA and Bosch. The complexity of all various subsystems and the effort for getting them running properly and safe demanded for further work structure before starting the whole process. This additional structure was called the *OpEneR Technology Levels* (left part of Fig. 2). Both the real integration of subsystems into the prototypes and the development of new functionality based on corresponding information sources were considered here. Some functional development and testing may already be able in the vehicle, other development work needs to be prepared and evaluated in pure simulation first and transferred to vehicle later.

Technology Level 1: All subsystems that are mandatory for a proper EV control (batteries, BMS, thermal management systems, charging system, etc.) are available as information sources and actors. Thereby a beyond state-of-the-art propulsion and energy recuperation through two electric motors plays an important role. High level recuperation is enabled by an innovative regenerative braking system ESP®*hev*. Vacuum need for brake force amplification is eliminated by means of the novel electro mechanical iBooster. A new HMI also including navigation system is developed and integrated to support the information and instruction of the driver. The installation of surround sensing systems like radar and video sensors is prepared. All together the E/E-architecture with all Electronic Control Units (ECU) is defined and the communication lines are set up.

Technology Level 2: The second level of OpEneR technology includes early version of ACC based on radar and video sensors. An advanced ESP®*hev* functionality is developed that maximises regenerative braking on roads with high friction coefficient, while safety especially on low-friction surfaces is always maintained. Via satellite navigation system an electronic horizon (eHorizon) is considered that gives information on the class of roads, 3D slopes, heights, curves

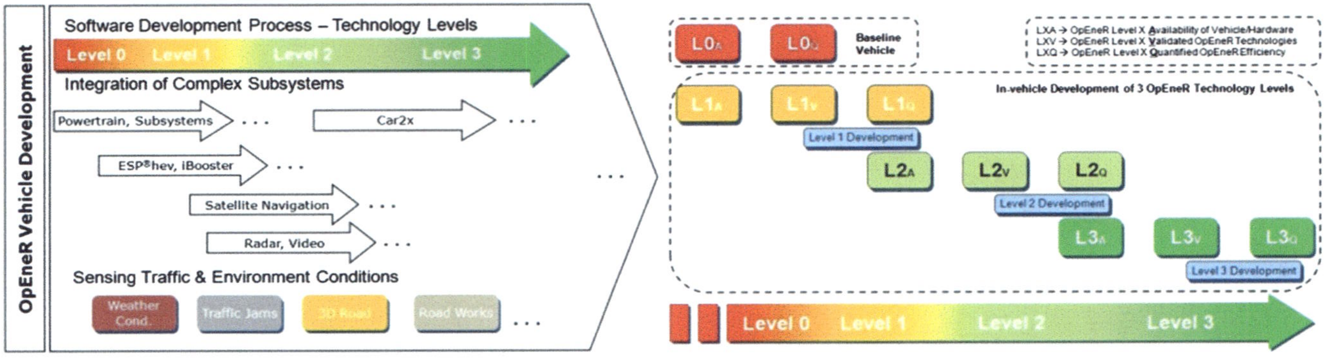

Fig. 2 OpEneR technology levels as structural measure to handle complexity of both the integration and the development process

and speed limits ahead. There is also a c2i technology included, e.g. for traffic light status information and road work positions.

Technology Level 3: The third and last level of OpEneR will be more speculative and is defined as final project progress. By means of additional usage of c2c technology, a more advanced economic ACC, more advanced road friction models from c2i or from on-board systems and a more sophisticated use of video camera information an additional benefit will be possible. In particular in this final level simulation plays a dominant role and real testing of functionality will be limited to some selected features only.

Following the development of the baseline vehicle in Level L0, the above mentioned three major OpEneR Technology Levels, L1–L3, are to be organised sufficiently well. Every Level LX (X = 1, 2, 3) is split in three phases ending with a milestone (see right part of Fig. 2). LX_A describes the milestone where the vehicle and additional hardware necessary for running level specific functions are both available. Milestone LX_V defines the status where level specific new functions are validated, milestone LX_Q the one where the benefit from new functions is quantified. There is an overlap between all four level specific development timelines.

2 OpEneR Prototypes

As mentioned above the OpEneR prototypes were planned to be build-up in the early project phase, in order to allow real testing of developed algorithms within the car as soon as possible. To achieve that and also to take the limited project budget into account, there was of course a need for a quick and cost reasonable decision for an existing vehicle model. After first specification and after discussion together with all partners and also with the Project Officer from the European Commission, PSA finally decided to go for the brand new Peugeot 3008 Diesel Hybrid4 (Hybrid4)—even not in series production at that time—as the basis vehicle. To reach the demand from the funding call it was decided to change the Hybrid4 into a fully electric vehicle by removing the Diesel combustion engine at the front axle and adding a second electric motor there. This process, also roughly depicted in Fig. 3, will be described in more detail in the upcoming sections.

2.1 Peugeot 3008 Diesel Hybrid4

The Hybrid4 vehicle is characterised by its novel axle-split drivetrain topology. Typically for most PSA vehicles, at the front axle there is a conventional combustion engine mounted. This strong Diesel engine delivers 120 kW power and a propulsive torque of 300 N m. On top a high voltage generator delivering 8 kW power at the front axle allows the Stop/Start functionality, i.e. the switch-off of the engine when stopping the car in front of a red traffic light and starting it again when

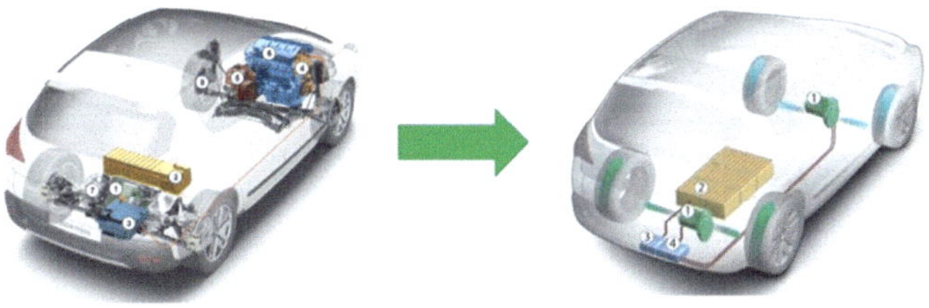

Fig. 3 Prototypes are evolved from the Peugeot 3008 Diesel Hybrid4

relieving the brake pedal. Although Stop/Start is a micro hybridisation approach already, the Hybrid4 goes a large step beyond and thereby runs a high voltage electric motor with 27 kW power and 200 N m torque at the rear axle (see left part of Fig. 3). Altogether, front and rear axle motors are not coupled.

The high voltage is delivered by a Ni-MH battery with 1.2 kWh energy capacity. Pure electric driving is possible for about 4 km and up to 60 kph. The control of the electric motors and of the DC/DC-conversion is handled by an all-in-one double inverter working on a voltage level between 200 and 270 V. Recuperation is supported by a so-called Regenerative Braking Control system (RBC) that helps to apply a combination of regenerative torque by the electric motor and a hydraulic torque by the friction brake at the same time. When only relieving the acceleration pedal but not yet braking, there is also a simulated drag toque provided by the electric motor that proves a deceleration of 1.2 m/s^2.

2.2 Prototype Build-Up

The electric axle-split architecture of the OpEneR prototypes is defined by two permanent magnet synchronous 50 kW mechanical power electric motors, one per axle. Thereby each electric motor is controlled by an individual inverter that is based on the reuse of the actual Hybrid4 powertrain (see right part of Fig. 3). Concerning electric motor and power electronics there were only slight modification of hardware to achieve maximum speed increase and power increase according to the new nominal voltage specification of 310 V. Both new inverter and vehicle control unit (VCU) software were necessarily developed in order to meet the open platform requirement. Thereby the main part of the original inverter software was transferred into the new VCU which is a dSpace MicroAutobox, and only low level control software was kept within the inverters.

To allow free rolling of the prototypes the electric motors are connected via an individual dog clutch that allows full decoupling. In addition a single stage reduction gear with transition ratio of 7.5 is used. Altogether the decisions were based on high degree of reuse of serial production systems and components.

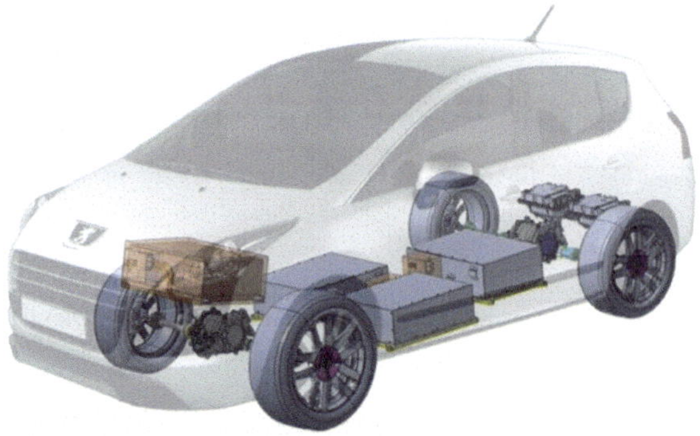

Fig. 4 Target configuration of the OpEneR prototypes

The voltage level of the OpEneR prototypes was defined by the maximum allowed for the current technology of Hybrid4 inverters, i.e. absolute 350 V maximum. The main drivers for this decision were build-up planning, packaging, costs of prototypes and also the operational safety. The design with four battery packs (see Fig. 4) was driven by cost, safety and technical reasons. Each pack is a 72 V autonomous battery package, which is a standard size for PSA's battery supplier. The usage of LiFePO4 battery technology was also driven by being a slightly better compromise in terms of safety and battery management, and since the ratio weight to capacity was not too critical. Basically, the batteries were chosen first, and the prototype was built around it. The four packs serial distribution is mainly driven by safety, since if one pack fails, the whole battery goes to safety. The cooling system was designed for battery needs to meet 150 km range and to achieve a behavior similar to serial production. There is a total battery capacity of 40 kWh delivered by 4 × 125 kg packages in serial connection. Usable capacity is 36.8 kWh (200 km range with respect to NEDC). Depending on temperature, the maximum power in charge as well as in discharge can reach up to 120 kW at a 3C rate, without compromise on safety or durability. The charging strategy is compliant with standard 16 A CEE plug in order to avoid extra time and budget for fast charging, and as well as to preserve the battery lifetime expectancy.

The thermal management is defined by two radiators for both sets of inverter, electric motor and battery. Water heating via 2 × 6 kW CTP is installed for cockpit and battery. In addition an electric A/C compressor with double circuit for battery and cockpit is integrated.

Concerning the body and the chassis design, major changes are needed in order to reach proper battery package integration (see Fig. 5). Keeping the same driver and passenger position, mounting seat points are lifted by about 15 cm in order to mount two battery packs underneath the main body frame helping keeping the centre of gravity as low as possible. The target for chassis modification and for body stiffness of OpEneR prototypes was to achieve a general behavior similar or better compared to the Hybrid4 although total mass is increased by 200 kg.

The electro hydraulic power steering was reused from the serial production to avoid expensive and high effort adaptations for an electric power steering.

Fig. 5 OpEneR prototype build-up in Paris in December 2011

2.3 OpEneR Subsystem Integration

The integration of new or modified subsystems into the first prototype started in February 2012. After the new brake system components were installed, the newly developed HMI components were mounted, including satellite navigation system and an active acceleration pedal. At the end of this section Table 2 summarises which partner contributes by which subsystems.

2.3.1 Braking System

Braking system components that are used in today's conventionally driven vehicles are normally a vacuum booster as pneumatic brake force amplifier and a standard ESP® system. The booster uses vacuum generated by throttle losses for gasoline engines or by a mechanical or electrical vacuum pump for Diesel engines. An ESP® system—including Anti Blocking System (ABS), Traction Control System (TCS)

Table 2 Suppliers of OpEneR subsystems

System	Delivered/produced by
Peugeot 3008 full electric, incl. battery and battery management system, thermal management	Peugeot Citroën Automobiles SA
Vehicle Control Unit (VCU)	Peugeot Citroën Automobiles SA
Electric motors and power electronics	Robert Bosch GmbH
Braking system components (iBooster and ESP® *hev*)	Robert Bosch GmbH
Surround sensing systems (radar and video)	Robert Bosch GmbH
Active acceleration pedal	Robert Bosch GmbH
Satellite navigation system	Robert Bosch Car Multimedia GmbH
Human Machine Interface (HMI)	CTAG—Centro Tecnológico de Automoción de Galicia
Car-to-Car and Car-to-Infrastructure communication units	CTAG—Centro Tecnológico de Automoción de Galicia

and Vehicle Dynamics Controller (VDC)—modulates the brake pressure for each wheel brake individually to enhance vehicle traction and stability. For hybrid electric vehicles and for EV the so-called iBooster provides electro mechanical brake force amplification (independent of vacuum) and the regenerative braking system ESP®*hev* supports recuperation via electric motors. Both systems are used within the OpEneR prototypes.

The ESP®*hev* system is characterized by a rear axle “brake by wire” circuit that is fully decoupled from the brake pedal during normal operation. In other words, driver brake pedal application does not directly result in a brake pressure increase on the rear axle brakes. Within a dead stroke in the tandem master cylinder (TMC) and based on the signal from a pedal stroke sensor, the braking demand given by the driver is covered by the electric motor regenerative braking only. Beyond this dead stroke, the brake pressure in the front axle brake circuit will be directly builtup via the master cylinder. The amount of brake torque at the rear axle is covered by the electric motors as long as the recuperation potential of the system is not exceeded, and no stability critical driving situation occurs. Since the recuperation potential of electric motors varies over vehicle velocity, the appropriate supplementary frictional brake pressure to achieve the desired vehicle deceleration is generated by the ESP®*hev* system within the rear axle brake circuit. This compensation is called *brake torque blending*. Here, the driver is not part of the control loop but experiences the well-known braking behavior like in conventionally driven vehicles. In case of a system degradation, e.g. due to component failures, or when a situation of critical vehicle stability occurs, the ESP®*hev* system is able to decrease or instantly and completely switch-off the brake torque from the electric motors in order to guarantee safe vehicle operation.

Shown over the different domains of actuation, power supply, modulation and foundation, the transition between conventional brake system components and the ones used in the OpEneR prototypes is visualized in Fig. 6.

2.3.2 HMI Including Navigation System

Advanced energy management functions that aim for most efficient driving strategies have influence on the usual driving behavior. Therefore the need for a reasonable and most flexible HMI is very high. Figure 7 shows on the left side the original HMI of the Hybrid4 together with the modifications that were undertaken. One of the major changes is the integration of a large and fully programmable TFT display replacing the conventional instrument cluster. Furthermore the gear-box selector is replaced by a knob selector and an iPad runs the new satellite navigation system. Other HMI specific systems like the original display in the middle of the dashboard or the audio system are just removed or rearranged.

State-of-the-art methods are used to define and to develop the new OpEneR HMI. The focus here lies on the separation between display of information during

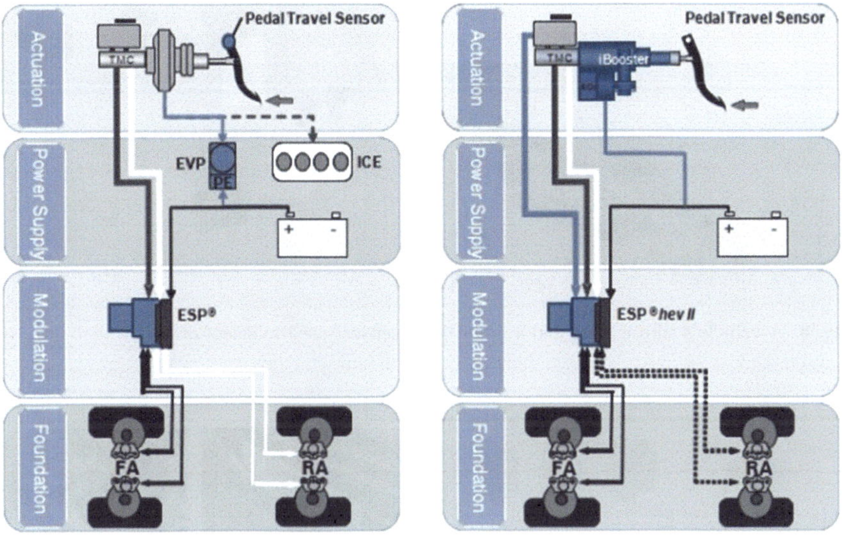

Fig. 6 Conventional brake system architecture on the *left side* and OpEneR EV brake system architecture on the *right side* [2, 3]

Fig. 7 HMI of series production Hybrid4 plus defined modifications, distribution of HMI functions and new OpEneR HMI

parking situation, information necessary for pre-trip, for on-trip and also for post-trip situation. Further distinction is put on vision information, control menus, feedback information and on external communication (see Fig. 7 in the middle).

The specification and development process finally results in the HMI depicted in the right side of Fig. 7.

2.3.3 Surround Sensing Systems

On the left side of Fig. 8 various sectors of surrounding environment of a vehicle are shown. Sensors that are widely used to detect objects within these areas are depicted in the right part of Fig. 8. Very near objects are observed by ultra sonic

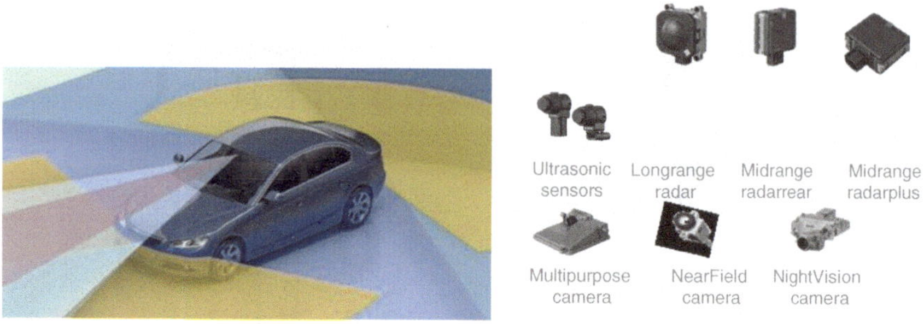

Fig. 8 A vehicle's surrounding and a portfolio of sensors used to detect objects

Fig. 9 Surround sensing systems in the OpEneR prototypes

sensors or near field cameras. Objects within larger distances are detected by radar sensors of different beamer range and by video cameras.

The information from such sensors is used to develop and to establish functions like parking assist and ACC. The OpEneR project aims for energy optimal driving strategies. The novel Eco ACC function therefore uses a mono video Multipurpose Camera, a Long Range Radar sensor and two Mid Range Radar sensors (Fig. 9).

2.3.4 Active Acceleration Pedal

An active acceleration pedal is installed within the prototype in order to support the driver with haptic feedback while driving. This is a quite efficient approach for informing and teaching the driver since the haptic channel usually is not such overloaded as for example the optical and the audio channel. Haptic feedback may be given as knocking, vibration, force feedback, variable kick-down or parallel force increase. Pedal functions can be triggered from different connected systems.

2.3.5 Car-to-Infrastructure and Car-to-Car Units

Finally c2i and c2c units will be mounted to the OpEneR prototypes. The target is to develop and to demonstrate functions that take benefit from information provided by surrounding and upcoming infrastructure and from other vehicles as well. In a final demonstration at the end of the project all functions are meant to be demonstrated on the so-called *Intelligent Corridor* in Spain that runs several RSU and provides server connection.

3 Energy Management System

Some functions of the OpEneR energy management system were already published (see e.g. [4, 5]). However, since the OpEneR project is still ongoing when preparing this article, functions and results of the energy management system cannot be fully described here.

All OpEneR partners are working jointly on the development of the energy management system that coordinates newly developed functions helping to reduce the energy consumption of EV. The whole process of development is supported by intensive simulation using a novel overall OpEneR simulation environment based on Matlab®/Simulink. The powertrain is modelled in AVL CRUISE and the vehicle is modelled within IPG CarMaker. Functions like regenerative braking, ABS, TCS and VDC can be simulated by means of the ESP®*hev* model integrated in this environment (see [6, 7,8]). Another simulation tool used within the OpEneR project is the so-called SystemC™ framework. This tool simulates data- and E/E-architectures and allows early evaluation of functional and non-functional behavior, e.g. communication latencies.

3.1 The OpEneR E/E-Architecture

Some new functions are implemented within specific electronic control units and others are implemented within the new VCU, i.e. a dSpace MicroAutobox. The whole E/E-architecture for hardware is visualised in Fig. 10. Due to requested quick prototype build-up many parts of the Hybrid4 E/E-architecture were reused. Major modifications were needed with respect to CAN HV (new powertrain and battery systems), to the Bosch CAN (new braking system and surround sensing systems), to the CAN I/S (sensor cluster and braking system) and finally to the HMI CAN (new HMI, navigation system, active acceleration pedal). Hereby state-of-the-art CAN bus technology was applied. The architecture is being intensively analysed in order to derive promising future modifications, e.g. concerning communication latencies. Due to time restrictions optimisations steps will not be realised in the OpEneR prototypes but described in a separate deliverable.

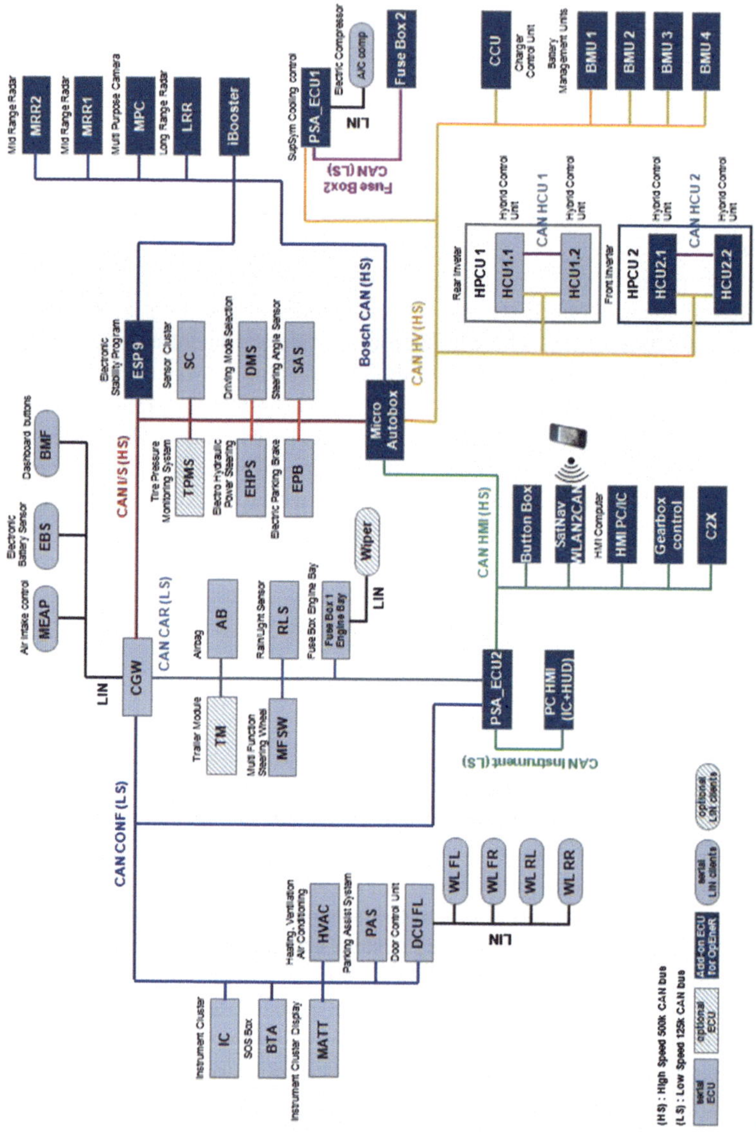

Fig. 10 OpEneR E/E-architecture with central VCU (dSpace MicroAutobox) connected to four main CAN buses

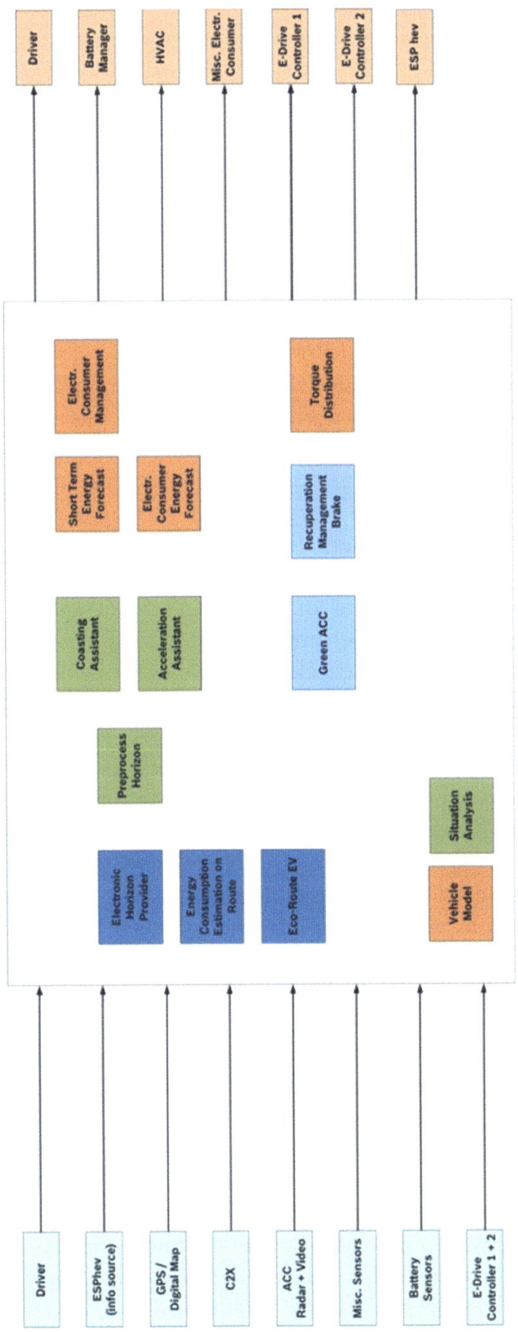

Fig. 11 Functional architecture of the OpEneR energy management system

3.2 Overview on Energy Management Functions

For the time being this article can only give an overview on newly developed functions that are part of the overall energy management system. A more complete picture together with results from all functions will be provided at the end of the OpEneR project by mid of 2014.

A high level description of the functional architecture is given in Fig. 11. Inputs for the energy management systems come from various information sources, e.g. the driver, the systems analysing vehicle situation and environment, and vehicle subsystems, like battery, powertrain and braking system. Outputs of the energy management system are being transferred back to the driver again (information and instructions), to the battery management system, to electrical consumers and to the actors that are part of the powertrain and the braking system.

Torque distribution: In general a 4-wheel electric drive given by the axle-split with two electric motors allows a better performance with respect to acceleration, vehicle dynamics and vehicle stability. This is particularly perceptible on low friction surfaces and in lateral dynamics situation. On the other side this drivetrain topology allows in the propulsive situation energy efficient shifting of drive torque between the front and the rear electric motor, always looking for the most efficient operation point of the combined system of two electric motors and two inverters. Thereby, in operating regions of higher total torque request a 50–50 split of torque is most suitable, but for smaller total torque demand a reduction down to only one electric motor is of benefit. In the case of deceleration also energy efficient torque distribution is possible. Here the efficiency of the battery system needs to be taken into account also.

Acceleration assistance: The knowledge of the efficiencies for the powertrain plus the functionality of the active acceleration pedal allow the implementation of an acceleration assistant that trains the driver to accelerate the vehicle in the most energy efficient way.

Regenerative braking: Next to drag torque simulated by the electric motors when relieving the acceleration pedal, high level of energy recovery via the two electric motors can be achieved by means of advanced software for the regenerative braking system ESP®*hev*. When braking via the brake pedal itself, the split of hydraulic brake torque and recuperative brake torque can be handled in a wide range of deceleration. In addition by communicating with both electric motors separately a shift of recuperative brake torque between the two electric motors can be achieved that helps to guarantee the installed brake force distribution of the vehicle. This supports the vehicle stability in most driving situation, independently of the specific road conditions.

Coasting assistance: Both the eHorizon (ADASIS interface) provided by the OpEneR satellite navigation system and the c2i communication unit can provide information on route sections with specific speed limits. When approaching such route sections the driver of the OpEneR vehicle is informed early to relieve the acceleration pedal and to allow free rolling. Thereby visual information on the demanded start of coasting, on the speed limit ahead and on its current distance are provided via a new HMI function. In addition the driver is informed haptically by a vibration of the active acceleration pedal to relieve the pedal. In the optimal case where all required data are accurate enough, the vehicle will reach its target speed at exactly the position of the detected speed limitation. In case of a foreseeable too high velocity at the position of speed limitation, from a defined position onwards also recuperation torque will be applied. When the coasting assistance took over, the driver is no more part of the control loop.

Eco ACC: When running an EV in the ACC mode, there are several differences to a conventionally driven vehicle to be taken into account. The most important difference of course is the availability of electric motors that give the possibility of recuperative braking. Normal friction brake with loss of kinetic energy can be partially or even completely avoided. Therefore the process of approaching other traffic objects, e.g. a vehicle ahead that runs with a slower velocity, can be handled in a more energy efficient way. Furthermore a cruise control that takes into account upcoming changes in speed limits will be implemented.

Eco routing and energy consumption estimation: The eHorizon with 3D map data together with traffic information provided by the satellite navigation system provides new ways of calculating the most energy efficient route between two points. An internal consumption model takes into account the recuperation potential of the electric powertrain and current consumption of other auxiliaries in order to predict the energy consumption on alternative routes.

HVAC conditioning: Finally the operation of auxiliaries and their energy consumption is taken into account. In particular for the energy demanding HVAC system an accurate consumption model is helpful for energy consumption modelling.

4 Conclusion and Outlook

The OpEneR prototypes and the newly developed energy management functions are being intensively tested during the last 2 years. Further functions are currently under development. Therefore testing will go on. With the achieved public road release for the OpEneR prototypes testing in real traffic conditions is now possible. Currently ongoing and thereby final steps of subsystem integration consider c2i and c2c communication units. Until the end of the project by mid of 2014 new energy saving functions using information from these technologies (e.g. traffic light assistance) will be developed and evaluated. An intensive testing of the overall final energy management system based on defined metrics will be done. Special emphasis will be given on the evaluation of aspects like modularity, scalability of the energy management architecture. As part of the third project review and a final event, all benefits from the work in OpEneR will be demonstrated on public road in Spain on the so-called *Intelligent Corridor*.

Acknowledgments The OpEneR project with grant agreement number 285526 is funded by the European Commission within the Seventh Framework Programme FP7-2011-ICT-GC. The authors want to thank all OpEneR project partners for significant contribution to the content described in this article.

References

1. OpEneR website: www.fp7-opener.eu
2. Quirant W, Kunz M, Hägele N (2012) ESP®*hev*: from vehicle stability to green braking—a modular approach to cover requirements for standard and electrified vehicles with one product family, chassis.tech plus 2012, vol 3. In: Internationales Münchner Fahrwerk-symposium, 2012, München
3. Willmann K (2012) From vehicle stability to green braking: actuation combined with modulation, intelligent braking 2012. In: IQPC International congress 2012, Wiesbaden
4. Köhler S, Viehl A, Bringmann O, Rosenstiel W (2012) Optimized recuperation strategy for (Hybrid) electric vehicles based on intelligent sensors. In: Proceedings of the 12th international conference on control, automation and systems, Jeju Island (Korea) 2012
5. Knoop M, Kern A (2013) Deceleration profiles for optimal recuperation and comfort. In: Proceedings of the 13th Stuttgart international symposium automotive and engine technology, Stuttgart 2013, vol 2, pp 85–98
6. Jones S, Kural E, Knödler K, Steinmann J (2013) Optimal energy efficiency, vehicle stability and safety on the OpEneR EV with electrified front and rear axles. In: Proceedings 17th advanced microsystems for automotive application, Berlin
7. Jones S, Kural E, Knödler K, Steinmann J, Braun V (2013) Simulated development of safe and energy efficient driving of the 4WD OpEneR EV with an advanced cooperative regenerative braking system. In: Proceedings of the 9th ITS European congress, Dublin, 2013
8. Jones S, Huss A, Kural E, Massoner A, Ludewig S, Knödler K, Steinmann J, Laversanne S (2013) Seamless development of vehicle energy management, recuperation and safety systems: pure office simulation to 4WD powertrain testbed. In: Virtual powertrain development conference 2013, Mainz

A Framework for Electric Vehicle Development: From Modelling to Engineering Through Real-World Data Analysis

Horst Pfluegl, Claudio Ricci, Laura Borgarello, Pacôme Magnin, Frank Sellier, Lorenzo Berzi, Marco Pierini, Carolien Mazal and Hellal Benzaoui

Abstract The Project ASTERICS, an international EU-STREP Project with 10 partners from 7 countries, follows the target to increase the efficiency of fully electric vehicles (FEV) by means of improved virtual models and intelligent testing and verification methods. Better models in the early design- and development

H. Pfluegl (✉)
AVL List GmbH, Hans-List-Platz 1, 8020 Graz, Austria
e-mail: horst.pfluegl@avl.com

C. Ricci · L. Borgarello
Centro Ricerche Fiat S.C.p.A., Strada Torino, 50, Orbassano, 10043 Turin, Italy
e-mail: claudio.ricci@crf.it

L. Borgarello
e-mail: laura.borgarello@crf.it

P. Magnin · F. Sellier
Siemens PLM Software, 84, Quai Charles de Gaulle, 69006 Lyon, France
e-mail: pacome.magnin@siemens.com

F. Sellier
e-mail: franck.sellier@siemens.com

L. Berzi · M. Pierini
Universita degli Studi Firenze, Piazza S. Marco, 4, 50121 Florence, Italy
e-mail: lorenzo.berzi@unifi.it

M. Pierini
e-mail: marco.pierini@unifi.it

C. Mazal
UNIRESEARCH, Unter den Linden 16, 10117 Berlin, Germany
e-mail: c.mazal@uniresearch-berlin.de

H. Benzaoui
Volvo Truck, Gropegårdsgatan, 40508 Gothenburg, Sweden
e-mail: hellal.benzaoui@volvo.com

B. Müller and G. Meyer (eds.), *Electric Vehicle Systems Architecture and Standardization Needs*, Lecture Notes in Mobility,
DOI 10.1007/978-3-319-13656-1_4

phases allow more realistic and improved concept studies and hence detailed optimization at component level as well as global optimization at system level. Through intelligent testing methods it will be possible to enhance the model quality on one hand and reduce the test time on the other hand. These testing methods shall also allow the assessment of durability and ageing effects for electrical components in the FEV-driveline. The combination of virtual simulation with realistic, for FEV relevant driving cycles leads to a very good possibility for optimization of predictable mileage. In this paper the results of the first phase of the ASTERICS project are presented. It describes the modelling approach and gives a good overview on virtual product development by means of model based system engineering (MBSE). Also described is the methodology to identify design goals based on real life data through assessment and definition of a representative driving cycle for FEV.

Keywords Battery · E-Motor · Inverter · Component model · System model · Advanced ageing · Advanced testing · E-Driving cycle · Predictable mileage · Electric vehicles · Realistic driving cycles · Electric powertrain simulation

1 General Approach

To facilitate the transition from conventional internal combustion engine (ICE) vehicles towards full electrically driven vehicles, the performance (range, speed, reliability, durability and efficiency) of Full Electric Vehicles (FEV) needs to be doubled whereas the cost of technology and development time has to be cut in half in comparison to state of the art.

There are currently quite a number of shortcomings in the design and development of FEVs. Simulation models of electrical components lack accuracy and interactions between components are not well known or explored. Real world optimization is lacking and FEV specific driving and stress situations not adequately taken into account. This leads to sub-optimized e-Drivelines and thus sub-optimized FEVs. Furthermore, testing procedures are ICE-optimized and efforts to adapt to FEV demands are unsatisfactory up to now. ASTERICS addresses this situation by taking a systematic and comprehensive approach for the design, development and testing phases of E-Drivelines for the FEV.

1.1 Concept and Objectives

The concept of the ASTERICS project comprises of four building blocks directly linked to the main objectives of the project:

1. Real world environment and conditions based drive cycles
   A set of real-world drive cycle usage profiles will be defined based on real-world use of the FEV, customer demands and fleet data [1, 2]. This information will be used to identify and specify representative operating conditions, performance requirements and constraints for FEV components, as well as stress scenarios, use cases and assessment criteria. To describe real-world contexts, different driving cycles will be provided, including urban/extra-urban/motorway cycles or specific applications, such as freight delivery and passenger public services.
2. Advanced testing methodologies and models for E-Driveline components
   Enhanced testing methodologies will be developed using Software, Model and Hardware in the Loop (SiL, HiL, MiL) environments and advanced models capable of simulating the real world behavior with the required accuracy and calculation speed. Standardized, accurate and time-efficient testing of E-Driveline components will be enabled in different phases of development and in different environments.
3. Electric sub-system models development
   Accurate high fidelity model for batteries, inverters and E-Motors will be developed considering all relevant dependencies from inputs/parameters, valid over the whole life-cycle of the E-Driveline components.
4. Total system (E-Driveline) models integration and validation on test bench
   The fourth building block addresses the complete system optimization integrating the high-fidelity sub-system models by means of components and interfaces, taking into account existing standards (like Modelisar-FMI). It will enable accurate simulation, calibration and optimization of the entire drive train in terms of vehicle performance and energy efficiency, and the usage/verification based on real world drive cycles in early stages of the development process.

1.2 Progress Beyond the State of the Art—Drive Cycles

Most existing driving cycles (both standardized/homologation cycles and research/industrial cycles) are conceived in order to evaluate performance and durability of vehicles in a repeatable way. The correct assessment of fuel consumption and air emissions usually are the priorities of such an approach. FEVs, on the other side, show a different sensibility to driving styles and environmental parameters. At the same time, the objectives of the analysis have to be significantly changed. It is necessary to switch the analysis considering different parameters, such as:

- The final energy consumption;
- The global duty cycle induced on the battery (e.g. including ageing during non-operating time);
- The influence of temperature on the performance;

- The interaction of the vehicle with the infrastructure, e.g. in terms of reduced warm-up phases in case of battery stand-by during charge, and the related energy consumption during such phase.

On the other side, conventional evaluations (e.g. gear shifting strategies) are not necessary (or have to be modified) to comply with the different driveline concept. Despite the fact that FEVs are already on the market, the know-how about FEV specific components durability is still limited. It is well known, however, that the instantaneous performance and the long-term reliability of critical components—such as the battery—are significantly related to the storage and operating temperature [3]. Also, the evaluation of other components performance—including power electronics—under combined temperature, power and speed effects is a challenging issue. Currently available driving cycles for E-Vehicles are not dedicated to the specific needs of electric vehicles with respect to urban, sub-urban and geographically demands.

Today, electric traction reaches new levels of power (>80 kW). Such power levels combined with high demanding drive cycles, which are the main target for FEV in urban drive cycles, require an accurate sizing of the powertrain. No other domain can provide feedback or data. Therefore, new methodologies and new tools for simulation, optimization and testing are required by OEMs.

ASTERICS will derive requirements, performance and energy efficiency demands specific to the needs of FEVs by identifying the given drive-cycles with real world data. Multiple scenarios will be proposed in order to fit with the complexity of the vehicle and the different mission profiles. The aim is to find new reference parameters reconsidering all the conditions that are currently accepted to develop real-world driving cycles and usage scenario (see Figs. 1 and 2).

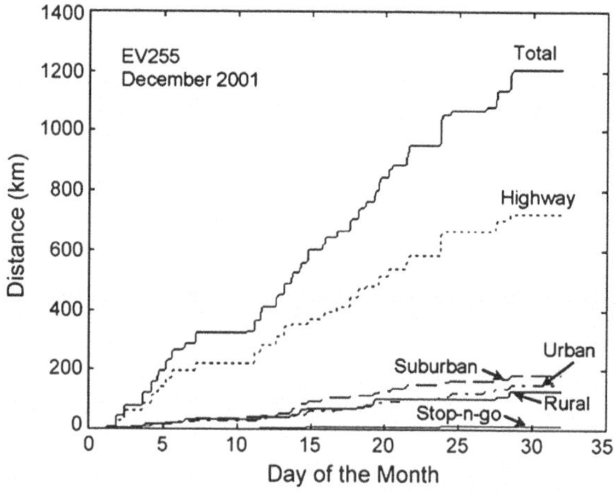

Fig. 1 Battery stress factor for BEV

Fig. 2 Usage classification for BEV

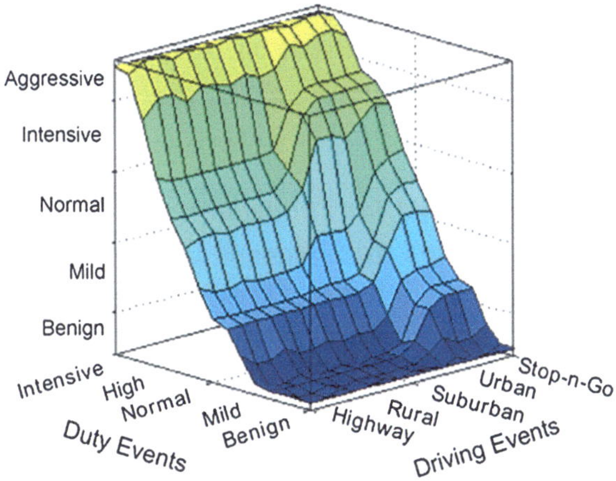

1.3 S/T Methodology and Work Plan

In the first phase of the project, requirements and specifications will be defined in order to propose a methodology able to identify the requirements of the FEV main vehicle components and the procedures needed for product development, testing and validation. Detailed use-case scenarios are built to be used during modelling and development phases. The accuracy of the results is important for the optimization of those components (both hardware and software) that have a critical impact on expected FEV performance, such as efficiency, energy consumption and resulting driving range. In three vertical activities a set of models will be developed for battery systems, inverter systems and the E-Motor. The use of these models in virtual and physical testing will be evaluated and the actual testing results provided. Subsequently, findings will be integrated into a complete FEV simulation model, including virtual driver setup and driving cycle description and numerical to experimental validation. A full ‘next generation’ vehicle model will be developed tuned on existing vehicles taken as reference. This simulation model will include all the components of a ‘next generation’ FEV: battery (with Battery Management System), e-Motor and inverter. On the experimental side, a set of roller bench test cycles will be created, with the intent to fit to the ‘next generation’ vehicles attitudes, taking into account climate conditions, driving style, and the traction archetype in study, leading to more reliable testing in the future.

2 Scopes and Potentialities of Modelling Activities

The use of software engineering tools enables the possibility to virtually design the system (product or process) under study and to virtually test—through simulations—the operational performance of existing or potential designs. The importance of

"Information Technology" tools in the engineering process has been constantly increased in the last 30 years; the whole approach to New Product Development (NPD) has been innovated.

2.1 The Impact for Innovation of Modelling Tools on Product and Process Development

Virtual simulation tools, at a first level, can act as a means or method to integrate or substitute physical experiments, main effect being the speed-up of the design process. As stated by some authors [1] this is only one of the advantages related to the extensive use of virtual simulation tools; their contribution can go well beyond the incremental improvement of the results obtained with physical experiments. The cited authors describe a significant impact of virtual simulation tools on R/D, resulting in a general modification of the design process: the need for formalization of Procedures for Virtual Development (PVD), which are aimed to effectively apply virtual tools in product/process development and to maximize their benefits. The roles of an efficient PVD should be, amongst others:

- To enable collecting of all necessary data on how product/process has to be designed
  - Design norms
  - Existing know-how
- To specify the whole flow of activities to be performed
  - Correct use of tools (how-to guidelines).
  - Simulation input/output analysis.
  - Implementation in practice (step by step).

As a first consequence, the described approach PVD can guide the whole design process, determining the interaction between components and the "incorporation" of all relevant data coming from company knowledge-base.

Therefore, as a second consequence, the PVD-based design process has an impact on the organization of the R/D team because of the need to coordinate different competencies and methodologies—namely formalized competencies, modelling and simulation capabilities and physical experiments to get data for validation. It can be highlighted the opportunity to include people coming from different scientific and experiential background in the same team, thus unifying physical and virtual simulation in the same unit.

Since one of the major advantages of virtual testing tools is the possibility to speed up testing at "relative" low cost, a higher number of alternatives can be verified in a given time. At the same time, if PVD provide correct standardization of the procedures, a stable base-line can be defined. The importance of base-line results is that the effect of each parameter variation can be compared with each other, enabling learning from earlier experiments. It is possible that a small amount of the results of virtual experiments is somehow unexpected (*puzzling* [1]): in such cases, a redefinition of the problem and of the search space can be proposed. In other words, a large number of virtual experiments also create opportunities for innovation. The goal of ASTERICS include the creation of a modelling framework for electric vehicle powertrain and the definition of the vehicle use-context starting from user-needs up to the proposal of testing procedures representing naturalistic ageing: all this elements can be part of a general PVD activity, in this case defined as Model Based System Engineering—MBSE.

2.2 Model Based System Engineering

The vision described above is substantially analogue to that proposed by other researches, which are describing a path for integration of CAE tools according to MBSE principles that is "the formalized application of modelling to support system requirements, design, analysis, verification and validation activities beginning in the conceptual design phase and continuing throughout development and later life cycle phases" (see [2, 3]). A tool aiming to support such process has therefore to cover following topics:

- Requirements
- Functional architecture information
- Logical architecture information
- Physical and software block
- Product configurations containing instances of physical elements, software and product parameters.

Each part interacts with other system parts. Cross phase activities establish relations between results coming from different phases, mainly being:

- Test Cases and Environment, that can be performed at different levels of abstraction
- Allocation and Analysis for requirements tracing, function partition and any kind of analysis.

Various approaches and software platforms can be used to build an MBSE environment, main need being flexibility, since each part of the system has to be defined according to its nature: datasets, executable codes, containers for mechanical and electrical properties, signals etc. Therefore, the suitability of candidate software and/or languages needs to be evaluated on the basis of MBSE needs; for a complete review see [4].

2.2.1 A General Modelling Framework

The MBSE methodology applied to vehicle design aims at finding the right vehicle component types, sizing and architecture, fulfilling design goals and allowing the realization of trade-off between design goals, formally analyzing the impact of these choices on the vehicle performance and behavior. Relying on physical component models, the MBSE design process brings better confidence on preliminary design, as it limits the risk of designing physically incoherent systems, leading to wrong specification, and later development issues.

A general framework for the creation of an MBSE methodology has been generated; the main steps composing such process are:

- Definition of a general dataset of boundary condition for vehicle design
- "Real life" data: driving cycles, charging cycles, environmental conditions and other
- Identification of vehicle design goal
- Explicit design goals
  - Regulatory goals: e.g. EURONCAP safety goals
  - Standard automotive goals: e.g. dynamics, comfort, performances
  - EV specific design goals: e.g. range, energy consumption expressed in appropriate indicators (Wh/km)
- Architecture design goals for, modularity, cost, use of existing components
  - Machine technology choice
  - Inverter technology choice
  - Battery chemistry and structure
- EV specific safety goals
  - Electrical insulation
  - Fire hazard in case of failures
- EV emerging design goals
  - For battery: energy density; ageing in terms of cell capacity dispersion and whole capacity decrease.
  - For electric motor: power density, torque, thermal issues, maximum torque depending on duty cycles.
  - Power electronics: temperature peak limit for ageing
- Definition of a simulation methodology
- Vehicle attribute validation through progressive steps and their iterations
  - Power sizing models
  - Energy sizing
  - Efficiency sizing

- Vehicle dynamics for comfort and performances
- Vehicle thermal comfort.

Figure 3 shows the approach for efficient MBSE methodology proposed within the first phase of ASTERICS activity. The flowchart explains the main steps of the proposed approach, its decision criteria and main inputs and outputs in this process, which are derived from the above listed general framework.

2.2.2 Guidelines for Model Creation

The last step of the proposed MBSE process aims to define "how to" progressively increase the detail of the model by incorporating improved input data, thus translating general requirements into detailed specifications. Such a process is critical because for its nature, it requires a number of models, having different origin, being executed together. The transition from one step to the subsequent one also requires increasing details of the models, but since iterations are possible, a general compatibility in terms of input, output and solver usability is needed. For this reason, a set of guidelines for the concept of the model for typical EV components has also been proposed starting from a comparison of the know-how of partners participating in the project. The minimum set of input and output ports as well as the usually needed parameters have been defined for:

- Battery pack: e.g. voltage, capacity, no. of series/parallel cells, parameters for grey box modelling (open circuit voltage, diffusive cell resistance, ohmic cell resistance, etc.), thermal data.
- Electric machine: e.g. technology, mass, rotor moment of inertia, reducer ratio, torque in motor/generator mode, max transient torque and power, base and maximum and operative speed at rated voltage, efficiency map/values…)
- Inverters and converters: mass, power, phase current, semiconductors modelling parameters (ON resistance, forward voltage-drop, switching frequency, switching energy…), efficiency map/values.

The use of a shared set for signal and physical ports is the first requisite for the integration of sub-models created independently from each other and used together in a common simulation environment. The second main need enabling the effectiveness of the simulation (in terms of acceptable ratio between results accuracy and simulation time) is the homogeneity of model complexity, which is mainly expressed as the minimum suitable time-step needed by the solver. These characteristics therefore, influences also the total simulation time. Main suitable model families depending on the typical goals of each design phase are summarized in Table 1.

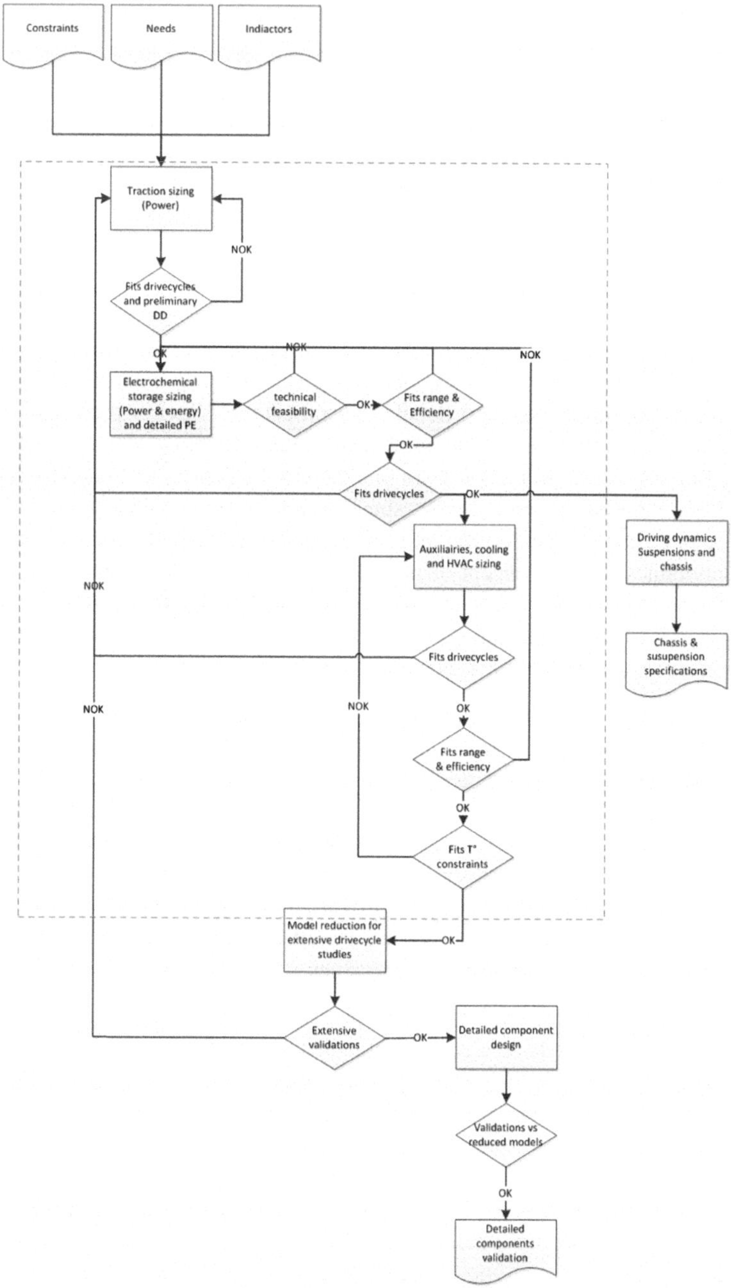

Fig. 3 First version of MBSE approach proposed within ASTERICS

Table 1 Modeling detail depending on design phase for main EV components

	Function	Element	Design phase			
			Archetype confrontation	Early design dimensioning	Design and testing	Component design
Components	Energy conversion	Motor	Efficiency table RPM/Torque performance	Simplified lumped parameters model	Component 0D–1D simulation through detailed lumped parameters models	Detailed FEM simulation
	Vehicle energy source	Battery fuel cell	Overall data kW/kg; kWh/kg; efficiency	Simplified model (e.g. source/resistor model)	Cells layout, thermal and electrical model	Detailed electrochemical simulation
	Power supply and control	Inverter/ controller	Overall efficiency	Suitable technology selection and expected performance	Component simulation (0D–1D)	Detailed simulation (voltage, current, EMC)
	Energy transmission	Driveline components	Overall efficiency	Preliminary dimensioning, expected inertia, stiffness	Driveline simulation (lumped models) (0D–1D)	Detailed multibody/FEM simulation
Typical solver time step			10^{-1}–10^{-2} s	10^{-2}–10^{-3} s	10^{-3}–10^{-4} s	10^{-5}–10^{-6} s

3 Integrating Real-Life Data in the Vehicle Development Process

In the described MBSE approach, realistic data is integrated in the design process, avoiding the drawbacks of designing electric vehicle fulfilling design goals driven by irrelevant regulation and mission profiles not coherent with the vehicle use. Above all, the most important data is the vehicle mission profile, made of, but not limited to:

- Realistic drive cycles (typically urban drive cycles for EVs)
  - "Standard" drive cycles: actual drive cycles
  - "Extended" drive cycles: integrate a slope profile
  - "Complete" drive cycles: integrate curve profiles
  - Charging cycles (specific to EVs)
  - Driver models (typical drivers behaviors)
  - Environmental conditions (average and extreme temperatures)
  - Vehicle loading

In general, a large number of factors are influencing the energy consumption of a vehicle, including driver behavior, driving context (city, countryside and highway), traffic conditions, and ambient temperature: such variability's are the reason determining the need for extensive testing on the road for any kind of vehicle during its final development phase. In order to reduce the cost of such critical phase, virtual testing and test-bed methodologies are used, and within this controlled contexts a "representative" driving cycle is needed as input.

A driving cycle can be considered as a standardized procedure aimed to evaluate vehicle performance in a reproducible way under laboratory conditions, such as simulation environment, power-adsorbing chassis dynamometer and/or component test-bed. It has to include a time-vehicle speed signal as main input data, but a large set of boundary conditions can be also defined: dynamometer settings, gear shifting points, reference atmospherics conditions (wind, humidity, pressure, temperature, etc.), vehicle conditions (tire pressure, lighting, oil viscosity, ...), "cold start" conditions (critical for different reasons, both for ICE and EV vehicles) and any parameter influencing the repeatability of the application. It is therefore clear that a single "driving cycle" cannot represent all the possible conditions under which the vehicle could be used during its entire lifecycle. Therefore some compromises in designing such a cycle have to be made. Considering EVs, the availability of a limited amount of on board energy is undoubted: under determined driving conditions, the loss of range (from about 150 km to about 50 % less) can determine a fail in satisfying customers need. Such issue is less critical on conventional vehicle due to a general larger range available (usually above 500 km) and quick fuel refilling options. For this reason, the EV's naturalistic mode of use has to be investigated, incorporating current fleet data as reference for future designs.

3.1 The Peculiarities of Electric Vehicles

Despite of the fact that the research on driving cycles started in the early '70s, the definition of naturalistic cycles and updated ones is still a topic under development in scientific and technical literature. According to a definition coming from an early work [5], "in general a time-speed history of "real" driving is selected from the assembled data in such a way that it matches the overall summary characteristics of the data set. This can be a contiguous sequence or part sequence of driving along a particular route, a selectively reduced longer trip along a particular route or a composite of randomly mixed and matched micro-trips from the survey data, usually involving some trial and error to achieve the correct statistical characteristics". Within the research activities, legislative driving cycles (e.g. NEDC) will still be considered for vehicle development; however, further effort for new cycle development is needed. A recent activity, based on an international collaboration, is aimed to defined a worldwide harmonized set of driving cycles (see [6]); within the present research activity, a specific study will be performed, according to modern driving cycle research methodologies (for a complete review, please refer to [7]) and to specific partners' know-how.

EVs introduce new parameters for the evaluation of their performance and are affected by specific, new criticalities in comparison to conventional vehicles. A brief list of such new factors includes:

- The possibility of energy recovery during braking, which could induce drivers to modify their style in order to optimize their energy consumption
- The limited range (about 150 km on optimal conditions for most EVs currently on the market) that could induce drivers to particularly smooth/benign driving style. Some specific conditions (e.g. occasionally high daily distances, unavailability of charging points, ...) are in particularly determining the so-called "range anxiety" phenomena
- A different "perception" of vehicle performances, due to different acoustic sensations, throttle characteristic, and torque availability from the powertrain
- An increased sensitiveness of the vehicle to the use of auxiliary systems.

The first aim of the research activity on driving cycles is to investigate the performances of electric vehicles and to identify if any peculiarity is recognizable in comparison with conventional vehicles; the second aim is the definition of a subset of driving cycles for performance evaluation.

3.1.1 A Methodology for Driving Cycle Analysis

The research activity defines a strategy for data acquisition and analysis. The process for the definition of a driving cycle is based on the following steps:

- Road/vehicle data acquisition in a predefined context
- Data pre-processing and preparation for analysis
  - Outliers examination
  - Filtering
- Data processing
  - Kinematic parameters evaluation
  - Speed-acceleration probability distribution analysis
  - Clustering
- Cycle synthesis
  - Boundary condition definition
  - Data selection/randomization
  - Verification
- Additional data proposal
  - Slope addiction
  - Consumption of auxiliaries
  - Environmental information

Within ASTERICS the acquisition of data from vehicles has been performed on more than one use-case in order to cover the various use-conditions for electric/hybrid vehicles components as good as possible.

In addition, the data—still under processing—will be analyzed by different partners, each one adapting the major indications of the procedure according to its know-how. The aim is to offer a number of vehicle information as an output in terms of kinematic parameters of acquired data (eventually grouped on the basis of their origin or using statistical tools), building up a comparison between/within different vehicles/context.

3.1.2 Case Study: City of Turin, Light Vehicle

The data acquisition on the city of Turin has been performed using a Light-Delivery-Vehicle (Iveco Daily, below 3.5t) running over a predetermined route comprising urban, extra urban and mixed urban roads. The vehicle has been tested in order to explore different conditions. The experiment plan includes the combination of various different parameters:

- 2 different professional drivers
- 2 driving styles (professional drivers requested to drive with different styles to obtain similar average speed in the mission)

- 2 loading conditions ("unloaded" and "80 % of max load")
- 2 different points in time during the day with different traffic conditions.

3.1.3 Case Study: City of Lyon, Heavy Vehicle

Data acquisition on the city of Lyon has been performed on a Heavy Vehicle (Volvo Premium Hybrid, equipped with an innovative hybrid powertrain, above 10t —GVW 19.2t). The vehicle is used in real condition in the city Lyon and province around it, so that there are no restrictive conditions about the daily path or about its distance. The vehicle is equipped with a data logging system for the powertrain and with a GPS system for geo-referencing, thus also data related to altitude variation will be included in the analysis. Due to the high mass of the vehicle, significantly different values are expected on parameters calculation in comparison with lighter ones, thus being well differentiated from the other case studies.

3.1.4 Case Study: City of Florence, Light Vehicles

Data acquisition on the city of Florence has been performed on a small number of light vehicles, and in particular:

- an electric quadricycle (Renault Twizy), mainly used within the area of the city for light freights delivery
- an electric van (Renault Kangoo ZE, below 3.5t class), mainly used within the area of the city for light freights delivery
- an electric passenger car (Peugeot iOn), used by various private and commercial users for general purpose trips.

Data are acquired during normal use condition of the vehicle. The main advantage of such approach is that data will be representative of the very particular use to which the vehicle are subjected, the city of Florence being affected by intense traffic within its historical center. After data acquisition, data will be stratified and compared with each other (in order to highlight eventually remarkable differences within the three main datasets); after that, a selection of data will be used for cycle synthesis, main aim being the characterization of the delivery activity in the specific service context and the comparison with general purpose use vehicles circulating in the same area.

3.2 Preliminary Results from Data Analysis

The activity of cycle development is still under study, especially regarding the processing of the data acquired. However, it is possible to set up a few

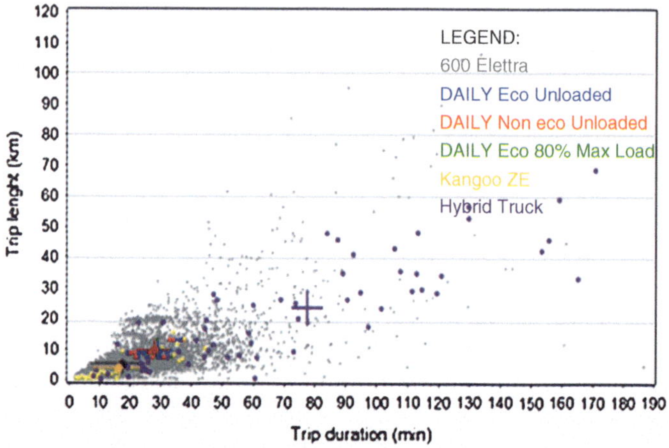

Fig. 4 *Dot-plot* showing trip duration and length for the dataset considered. Distances above 120 km, have been excluded. *Cross symbols* indicate the centroid of the distribution. Values are compared to a larger database coming from former ATENA project, using a Fiat 600 electric vehicle [8]

confrontations on the basis of partial data. About 400 km have been considered for Iveco Daily, about 300 km for Renault Kangoo ZE and about 1,600 km for the Hybrid Heavy vehicle proposed by Volvo.

3.2.1 Comparison of the Main Parameters

At the present time, it is possible to show the difference in terms of *trip* (a driving event included between key-on and key-off) duration and length. Figure 4 shows the results for the fleet considered and, in case of Iveco Daily, for its main use variants; the dots indicate that, as expected, different points are expected between very different missions: from short urban trips (considered for light van Kangoo electric), to longer trips in case of the heavy hybrid truck. Figure 5 shows the important parameter of mean positive and negative acceleration, thus defining the main driver "attitude" for each vehicle. Of course, different vehicles tend to fall in different sub-areas.

3.2.2 Synthesis of New Cycles

The synthesis of compression of driving data into cycles usually is composed by three main phases:

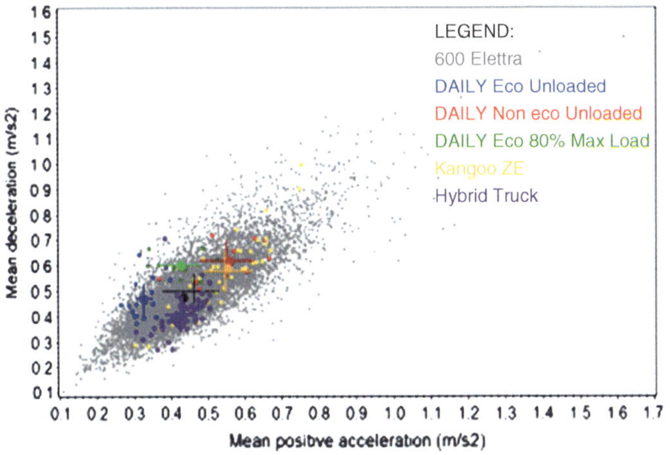

Fig. 5 *Dot-plot* showing mean positive acceleration and mean positive deceleration for the dataset considered. *Cross symbols* indicate the centroid of the distribution. Values are compared to a larger database coming from former ATENA project, using a Fiat 600 electric vehicle [8]

- According to the available database of driving sequences and of the associated kinematic parameters (see former paragraph), a number of mini-trips (speed signal between two stop phases) or other kind of segments are randomly selected according to the desired characteristics of the cycles (representativeness of proportion for each class)
- Sequences are "glued" according to "matching" criteria
  - coherence in terms of final speed of the preceding segment to the speed of its next segment; in case of mini-trips, speed is zero
  - once the target duration has been reached, a verification of the representativeness of the cycle is performed.

Within the research activity, data processing is still under development and, in particular, the comparison within each dataset (different for vehicle, context and mission) has to be performed. However, the preliminary result that is shown in Table 2 is a comparison between the original set of naturalistic driving data (column "Mean of trip" and "Global value") and a draft version for a synthetic driving cycle (column "Proposed Cycle" in Table 2)—as shown in Fig. 6, currently proposed for the Turin case study. The corresponding deviations in comparison to the "Proposed Cycle" are listed in Table 2 as Difference in %.

Table 2 Comparison of kinematic parameters between original data and proposed synthetic cycle (case study: Turin)

Category	Statistic	Proposed cycle	Mean of trip	Difference (%)	Global value	Difference (%)
Duration	Trip duration (min)	26.3	28.9	−9		
	Trip length (km)	8.7	9.6	−10		
Speed	Average speed (km/h)	19.8	20.1	−2	20.0	−1
	Stop time (%)	29.6	29.5	0	30.1	−2
	Number of stop at km	2.2	1.9	15	1.8	21
	Speed (km/h) without considering speed = 0	28.1	28.4	−1	28.6	−2
	95 ‰ speed (km/h)	42.7	46.7	−9	49.1	−13
Acceleration	Percentile constant speed (% moving time)	22.1	24.4	−9	24.4	−9
	Standard deviation in acceleration (with sign) m/s^2	0.53	0.52	1	0.52	0
	Mean positive acceleration (m/s^2)	0.34	0.33	6	0.33	5
	Mean deceleration (m/s^2)	0.45	0.44	2	0.44	4
	95 ‰ acceleration (m/s^2)	1.05	1.01	4	1.02	3
	95 ‰ deceleration (m/s^2)	1.02	1.20	−15	1.24	−18
	Initial acceleration (m/s^2)	1.19	1.21	−2	1.20	−1

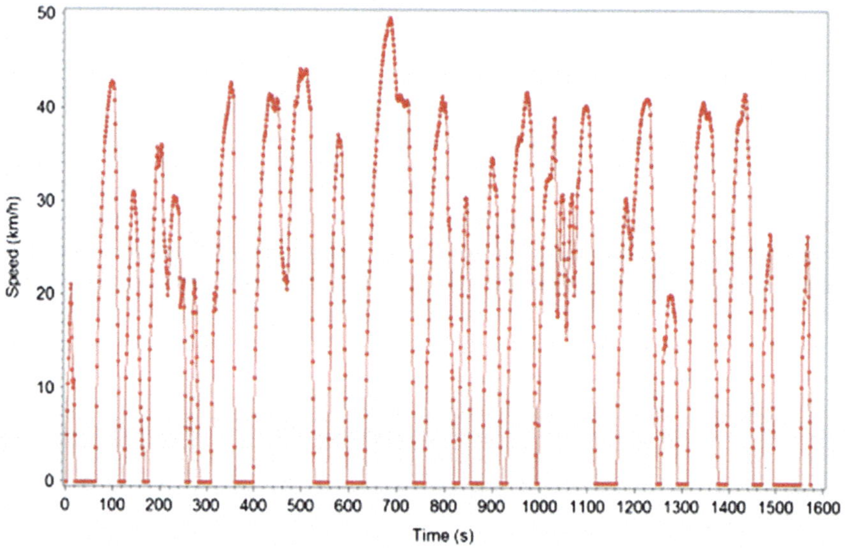

Fig. 6 Preliminary synthetic, real-world driving cycle derived from Turin case study

4 Conclusions and Outlook

The interim results in this paper have clearly shown that the virtual product development process (PVD) here also introduced as model based system engineering (MBSE) is absolutely necessary for the development of competitive electric vehicles. The relevant data and results of assessment from real driver experiments can be fed back to the development process in a much better and faster way. Models can be improved in this iterative approach and lead to better simulation results in the next iteration.

The investigations on the driving cycle and calculation of a proposed driving cycle are still ongoing in this project. It is currently possible to create realistic driving cycles for dedicated areas and dedicated usage scenarios. The impact to the design, efficiency and predictive mileage is subject to further investigations and will be shown later in this project.

Further research work is foreseen in the creation of realistic models which are able to reflect parameter changes in various combinations, including ageing behavior of most components of the E-Powertrain. At the end of this project, submodels will be linked together and simulated in combined environments and validated against measurement results, using the realistic usage scenarios that have been described in this paper.

References

1. Becker MC, Salvatore P, Zirpoli F (2005) The impact of virtual simulation tools on problem-solving and new product development organization. Res Policy 34:1305–1321
2. Votintseva A, Witschel P, Goedecke A (2011) Analysis of a complex system for electrical mobility using a model-based engineering approach focusing on simulation. Procedia Computer Sci 6:57–62
3. International Council on Systems Engineering (INCOSE): Systems Engineering Vision 2020, Document No. INCOSE-TP-2004-004-02, Version 2.03 (2007). http://www.incose.org/ProductsPubs/pdf/SEVision2020_20071003_v2_03.pdf
4. Fotso AB, Rettberg A (2012) State of the art for mechatronic design concepts. In: 2012 IEEE/ASME International conference on mechatronics and embedded systems and applications (MESA). Presented at the 2012 IEEE/ASME international conference on mechatronics and embedded systems and applications (MESA), pp 232–240
5. Lyons TJ, Kenworthy JR, Austin PI, Newman PWG (1986) The development of a driving cycle for fuel consumption and emissions evaluation. Transp Res Part A: Gen 20:447–462
6. UNECE (2013) Worldwide harmonized Light vehicles Test Procedure (WLTP). https://www2.unece.org/wiki/pages/viewpage.action?pageId=2523179 (last seen September 2013)
7. Tong HY, Hung WT (2010) A framework for developing driving cycles with on-road driving data. Transp Rev 30:589–615. Published Project Report PPR354. http://www.trl.co.uk (last seen September 2013)
8. Gortan L et al (2001) Preliminary results on emission and driving behaviour of ATENA fleet test project in Naples. In: Proceedings of the 5th international conference on internal combustion engines, pp 52–60

HiWi Project: High Efficiency Electric Drives

Andrew Cockburn, Jenny Wang, David Hopkinson, Marco Ottella, Fabrice Marion and William O'Neill

Abstract Vehicles develop their highest efficiency of around 93–95 % within a speed range of usually 1/4 to 1/3 of the maximum, whereas in real-life driving cycles the motor operates at a wider range of speeds and at partial load, resulting in much lower efficiency. Hi-Wi addresses this mismatch by advancing the design and manufacture of drive trains through holistic design across magnetic, thermal, mechanical and control electronics/algorithms in line with real-life use rather than a single-point "rating". In addition to the above efficiency gains, Hi-Wi addresses the

---

Project Partners University of Cambridge (UK), University of Sheffield (UK), Istituto P.M. SRL (I), Centro Ricerche Fiat (I), STMicroelectronics (I), CEDRAT SA (F), Siemens AG (D).

A. Cockburn (✉)
Institute for Manufacturing, University of Cambridge, The Old Schools, Trinity Ln, Cambridge CB2 1TN, UK
e-mail: ac282@cam.ac.uk

J. Wang
Health Economics and Decision Science, ScHARR, The University of Sheffield, 30 Regent Street, Sheffield S1 4DA, UK
e-mail: j.w.wang@shef.ac.uk

D. Hopkinson
Industrial Photonics, University of Cambridge, The Old Schools, Trinity Ln, Cambridge CB2 1TN, UK
e-mail: dmh48@cam.ac.uk

M. Ottella
Centro Ricerche FIAT SCPA, Strada Torino 50, 10043 Orbassano, Italy
e-mail: marco.ottella@crf.it

F. Marion
CEDRAT SA, 59 Chemin du Vieux Chêne—Inovallée, Meylan Cedex 38246, France
e-mail: fabrice.marion@cedrat.com

W. O'Neill
University of Cambridge, The Old Schools, Trinity Ln, Cambridge CB2 1TN, UK
e-mail: wo207@eng.cam.ac.uk

B. Müller and G. Meyer (eds.), *Electric Vehicle Systems Architecture and Standardization Needs*, Lecture Notes in Mobility,
DOI 10.1007/978-3-319-13656-1_5

issue of material supply through the development of nanostructured magnetic materials and the development of new driving cycles to more accurately represent in use conditions for electric vehicles.

Keywords Permanent magnet machine · Rare earth magnets · NEDC drive cycle

1 Introduction

At present, motors for FEV (Fully Electric Vehicle) and HEV (Hybrid Electric Vehicle) applications develop their highest efficiency of around 93–95 % within a speed range of typically 1/4 to 1/3 of the maximum rotating speed, and at an ideal torque, whereas in real usage—in the majority of driving cycles—the motor operates at a wider range of speeds and at partial load (low torque) resulting in lower overall efficiency, Fig. 1.

Hi-Wi addresses the mismatch between the region of HIGH efficiency and the WIDE region of frequent operation with advances in the design and manufacture of drivetrains optimized for performance over a whole drive cycle rather than a single point.

In addition to efficiency gains, Hi-Wi couples its novel design approach to developments in magnetic materials which utilize nano-scale microstructural control to deliver reductions in the consumption of rare earth elements. This is of interest due to the unstable nature of their supply chain and the rapid increase in price which has been associated with the increase in use of these materials, see Fig. 2.

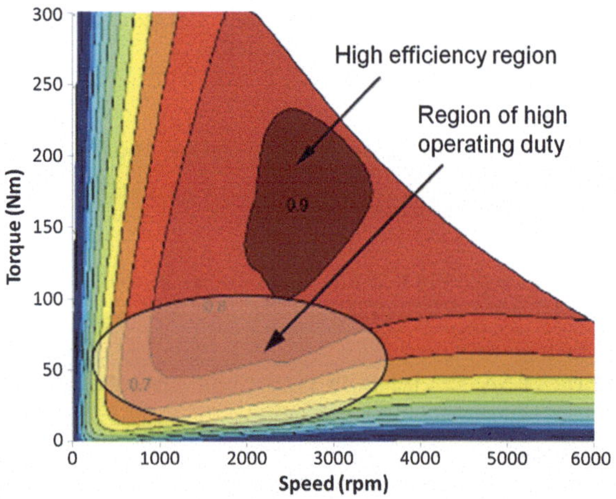

Fig. 1 Mismatch of high efficiency region and frequently operating region of conventional traction drives

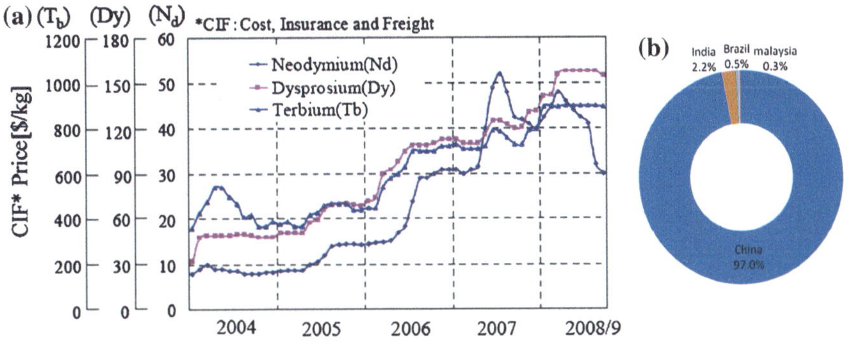

Fig. 2 Data showing the increase in the price (**a**) and concentration of supply (**b**) of rare earth elements [1]

The 3-year Hi-Wi project is set to deliver new design tools and configuration for an electric drivetrain optimized against a real driving cycle and novel process routes for magnet production which reduce the consumption of rare earth elements.

This paper considers the approaches taken to arrive at an optimal configuration for the drivetrain of a small electric vehicle and examines new approaches to the reduction of rare earth element use in high performance permanent magnets.

2 Analysis of Drive Cycles

One of the major themes of the project is the evaluation and optimization of the performance of the drivetrain throughout a real drive cycle. The cycles considered in this work were the New European Driving Cycle (NEDC) and the Artemis urban driving cycle [2]. NEDC has longer high speed segments with less demand on high torque and acceleration. As such a significant portion of the energy is consumed in high speeds when the required traction torque is relatively low. The Artemis urban, on the other hand, requires frequent starts/stops which are the characteristics of city driving. The peak acceleration requirement is also much greater than NEDC. Thus unlike NEDC, most energy consumption over the Artemis urban cycle is in the low speed and high torque region.

In order to evaluate the torque/speed requirements for the drivetrain, against drive cycles, the basic parameters of the proposed vehicle were defined. These factors, shown in Table 1, allowed the accelerations and speeds imposed by the NEDC, Fig. 3a, and ARTEMIS drive cycles to be converted into the power required as a function of time, Fig. 3b.

Given the power and speed profile dictated by the NEDC profile and vehicle parameters the distribution of energy consumption with torque and speed has been determined, Fig. 4.

Table 1 Vehicle parameters for cycle simulation

Vehicle parameter	Unit	SoA vehicle
Curb weight	[kg]	800
Test load	[kg]	100
Vehicle translating apparent mass coefficient	[–]	1.03
Tyre size		145/65 R15
Wheel rolling radius	[m]	0.284
Rolling resistance coefficients		
• f_0	[–]	0.008
• f_2	$[1/(km/h)^2]$	3×10^{-7}
Aerodynamic resistance		
• Frontal area	$[m^2]$	1.8
• Aerodynamic coefficient	[–]	0.25
Transmission		
• Fixed speed solution		
• Efficiency	[–]	0.93
• Reduction ratio	[–]	4 or 7

It can be seen that the energy is distributed over a wide torque and speed range. However, there are 6 points at which the energy consumption is much high than others. These points correspond to the constant speed operations of the NEDC cycle in which the time duration is much longer while the torque is relatively low. The motor efficiency over these points are therefore of great importance.

The characteristics of the energy distribution over the NEDC illustrates that in order to achieve high energy efficiency, the motor must be designed to have high efficiency over the 12 representative points. If this is not completely possible, then attention should be paid to have high efficiency in low torque and high speed range.

It is therefore indispensable to emphasize the significance of optimizing a traction machine over its corresponding driving cycle for maximum efficiency or minimum energy loss. An important point to note is the time needed for the finite element (FE) calculations of the motor efficiency over the driving cycle during the design process. In fact, with the current state-of-art computing and design tools, several months would be required in order to obtain FE predicted efficiency results for the hundreds of points over the NEDC. This would make it impossible to perform design optimisation process in a realistic time scale with limited human resources. However, this limitation is effectively removed using the twelve representative points already presented. To validate the above representation, the energy loss over the NEDC cycle is calculated by all the points in Fig. 5 using the efficiency map of a traction motor developed for EV traction of similar specifications.

Table 2 shows that the 12 points described in Fig. 6 can be used as an acceptable replacement for the entire drive cycle when optimising a drivetrain for maximum efficiency over the NEDC cycle.

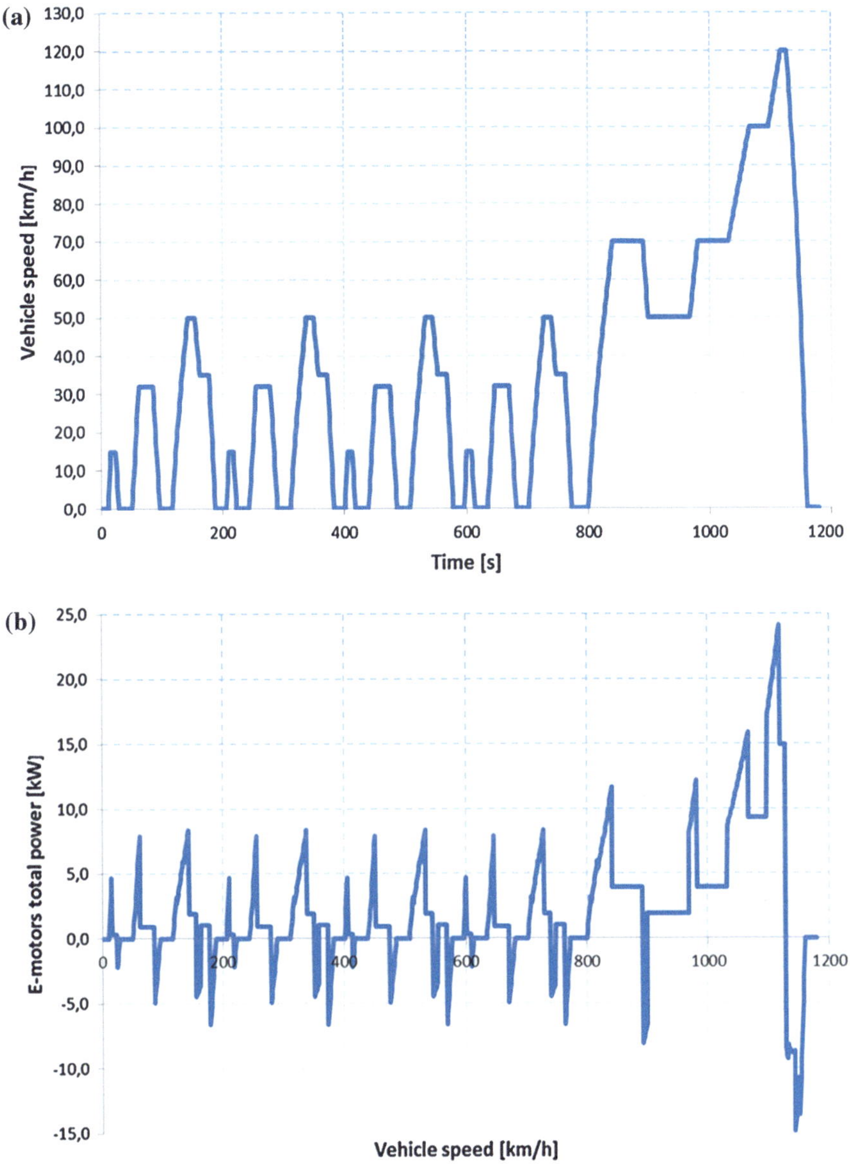

Fig. 3 Speed (**a**) and power demand (**b**) during the NEDC cycle

A similar approach has been taken with the Artemis urban drive cycle. This cycle is more heavily weighted to low speed/high torque operating conditions so that optimising given motor design for the NEDC cycle may reduce the efficiency of the system over the Artemis cycle.

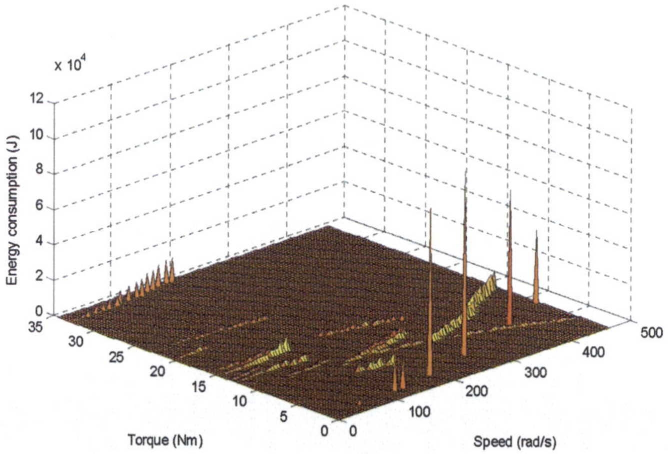

Fig. 4 Variation energy consumption with torque and speed over the NEDC drive cycle. This concentration of energy consumption allows the NEDC cycle to be represented by 12 points on a torque versus speed plot as shown in Fig. 5

Fig. 5 Distribution of 12 representative energy consumption points

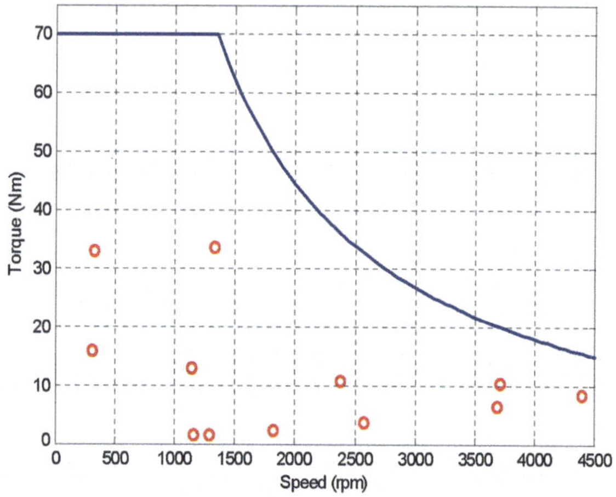

Table 2 Comparison of energy loss

Energy loss over all points in NEDC cycle (kJ)	55.104
Energy loss over 12 points in Fig. 6 (kJ)	53.387
Difference (%)	3.1

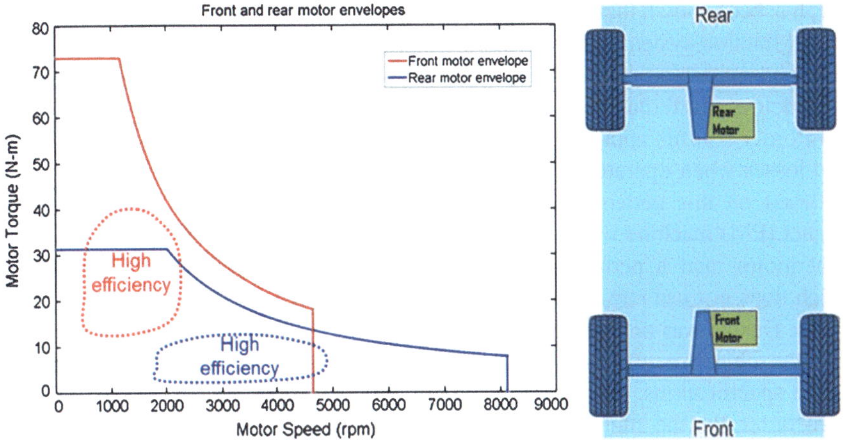

Fig. 6 Co-working envelopes of front and rear motors with different efficiency performances

3 Powertrain Design and Simulation

Candidate topologies tested as part of system for HiWi Principles and compared with constant torque split results. HiWi optimizes over NEDC and artemis.

Machine topology selection was based on a defined set of figures of merits for the most promising candidate machines to achieve high efficiency over a wide operating range. Parameters which were considered included torque density, power density, torque ripple, efficiency, controllability, reliability maturity, cost, manufacturability and fault tolerance.

Using this approach, the most promising topologies were found to be Stator Permanent Magnet Machines (SPM) and PM assisted Synchronous Reluctance Machine (Syn-RM).

These motor topologies were considered for use in the HiWi and optimized for use over the NEDC and artemis drive cycles using the vehicle data defined in Table 1 which impose the following parameters on the motors, see Table 3.

Table 3 Machine specifications and geometry constraints for HiWi front motor

Parameter (unit)	Value
Base speed (rpm)	1,350
Maximum speed (rpm)	5,050
Peak torque below and at base speed (N m)	70.0
Continuous torque below and at base speed (N m)	35.5
Peak torque at maximum cruise speed (N m)	15.0
Peak power (kW)	9.9
Continuous power (kW)	5.0

It has been shown that by employing a dynamic torque apportioning strategy, the overall traction system efficiency with distributed power train configuration can be improved with dissimilar machine technologies or topologies [3] giving peak efficiencies in different parts of the torque speed envelope, Fig. 6.

In order for this approach to be effective, the motors must have minimal low-load losses when operating outside their range of efficiency.

Based on this understanding, the newly developed fractional slot permanent magnet (PM) machine with low space harmonic contents has been selected for the front motor, and a permanent magnet assisted synchronous reluctance machine which does not use rare earth magnets was adopted for the rear motor.

The Design was optimised against the NEDC and Artemis Urban driving cycles, with the objective to minimise the energy consumption while satisfying performance specifications, electric, thermal, mechanical and volumetric constraints.

In order for the motor designs and drivetrain configuration to be optimized against real driving cycles, electromagnetic-thermal coupled field modeling and motor design tools developed by project partner CEDRAT were employed. Their applications in motor design illustrated in this paper. A computationally efficient design optimisation method has been developed based on in-depth understanding of the energy consumption of the vehicle over the reference NEDC driving cycle [4].

This together with an FE based optimisation tool (Got-It) enables the global optimization of the traction machine aimed at maximum energy efficiency over the driving cycle being performed in a very computationally efficient manner.

Since the NEDC and Artemis cycles represent differing loading cases, optimization has taken place over the NEDC and Artemis cycles and a combination of the two. The results of applying a dynamically varying torque split are displayed in Table 4.

Divergent efficiency trends emanating from the distinct characters of the driving cycles are observed for the rear wheel PMa-SynRM topology. Therefore, the performance of the machines optimised against NEDC and Artemis is penalised when they are operated in a driving cycle other than their optimisation target. However, the machine optimised against the combination of both cycles manages to achieve relatively high efficiency over both NEDC and Artemis cycles, see Table 5.

Table 4 NEDC efficiencies from different motor topology combinations with fixed and optimal power split ratios

Front motor	Rear motor	Power split ratio	NEDC efficiency (%)
12 s 14p SPM	12 s 14p SPM	0.5:0.5	94.39
PMa-SynRM	PMa-SynRM	0.5:0.5	93.70
12 s 14p SPM	PMa-SynRM	0.5:0.5	*94.01*
12 s 14p SPM	PMa-SynRM	*Optimal*	*94.94*

Table 5 NEDC efficiencies from different motor topology combinations with fixed and optimal power split ratios

Drive system cycle efficiencies	$\mathrm{Eff}_{\mathrm{NEDC}}$ (%)	$\mathrm{Eff}_{\mathrm{Artemis}}$ (%)	$\mathrm{Eff}_{\mathrm{combined}}$ (%)	PM usage (%)
Optimised for NEDC	*90.73*	−0.79	−0.10	100
Optimised for Artemis	−0.93	*88.82*	−0.35	+28.50
Optimised for combined	−0.23	−0.28	*89.60*	+9.39

4 Motor Manufacturing and Integration

Motor designs optimized for efficiency across a combined NEDC and Artemis urban drive cycle have been designed and manufactured to allow testing and integration into the WIDE-MOB vehicle.

4.1 HiWi Motor Designs

The machine topology is shown in Fig. 8.

It is an interior mounted permanent magnet machine with 18 stator slots and 8 rotor poles. The coil sides of the 3-phase windings are placed in the stator slots in superposition. The magnets are embedded in the rotor lamination in a V-shape to enhance torque production capability through the reluctance torque and to improve field weakening capability for high torque and wide operating speed range.

The efficiency map over the torque-speed operating envelope is shown in Fig. 7. The efficiency from 94 to 97 % can be achieved over a wide region with the torque up to 50 N m, the speed up to 5,000 rpm.

Optimisation studies described above suggested that a Permanent Magnet Assisted Synchronous Reluctance Motor (PMa-SynRM) was highly suited for the rear motor of the HIWI vehicle. The presence of the low-cost ferrite magnets in the

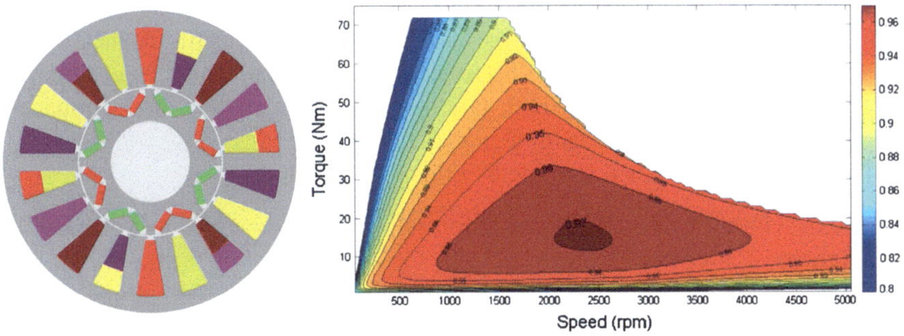

Fig. 7 Topology and efficiency map for front motor permanent magnet machine

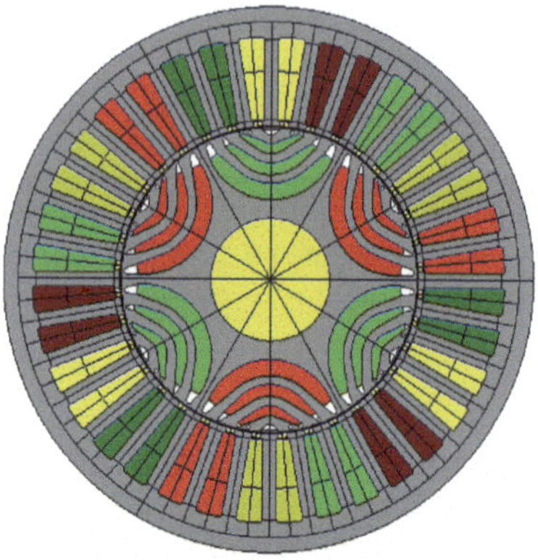

Fig. 8 Topology for rear motor permanent magnet assisted synchronous reluctance machine

rotor allows this topology to achieve a higher power factor and torque density compared to a purely synchronous reluctance machine (SynRM). A 2-D cross section of the PMa-SynRM is illustrated in Fig. 8.

For building two demonstrators, the tooling cost for ferrite magnets were beyond the available budget. Therefore, the motor is designed with bonded NdFeB with comparable remanence to that of ferrite magnets available on the market.

Figure 9 shows the test motors which are undergoing assessment prior to integration into the WIDE-MOB/P MOB vehicle shown in Fig. 10. WIDE-MOB and P-MOB are FP7 funded projects which have developed a vehicle to address the demand for urban mobility. The vehicle developed by the projects is a 3 seat EV which can sit in either the micro e-car (as defined by EU regulation 168/2013) or, with modification, the city e-car category. The car has front and rear air cooled motors and inverters, a safety cell capable of meeting NCAP test rating, is designed to mitigate electromagnetic fields and incorporates photovoltaic cells to provide up to 20 km/day of range.

5 Magnetic Material Manufacturing

One of the most pressing issues for the widespread implementation of electric vehicles using permanent magnet drives is the supply of rare earth magnets. Work on magnetic material within the HiWi project has concentrated on the development of process routes to allow the bulk production of high performance permanent magnets with a reduced rare earth element content.

This work has examined two routes to rare earth reduction:

1. The development of a process route for bulk exchange coupled nano-composite magnets
2. Control over the distribution of dysprosium additions to NdFeB permanent magnets to allow improved high temperature performance with reduced Dy content.

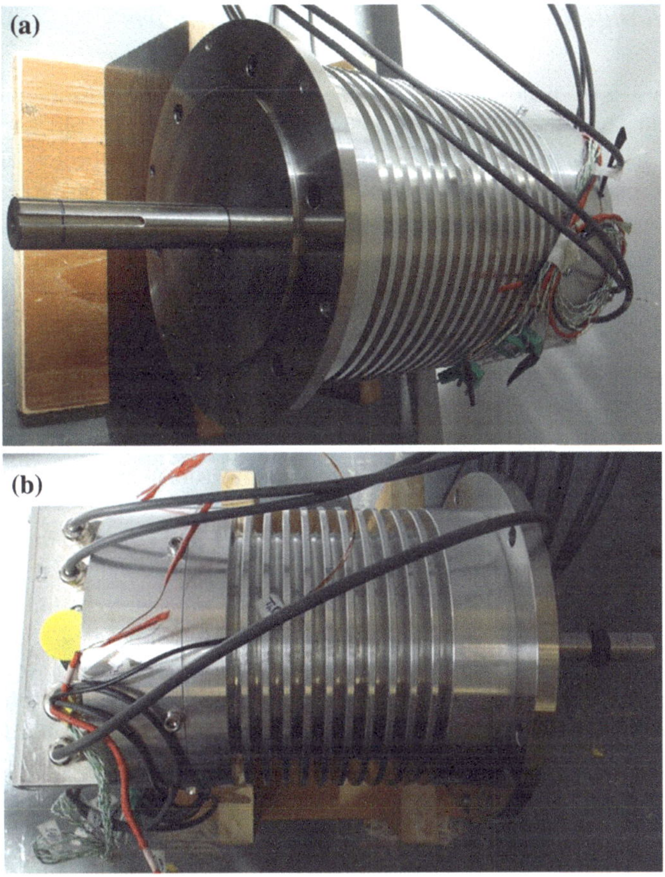

Fig. 9 Front (**a**) and rear (**b**) motors

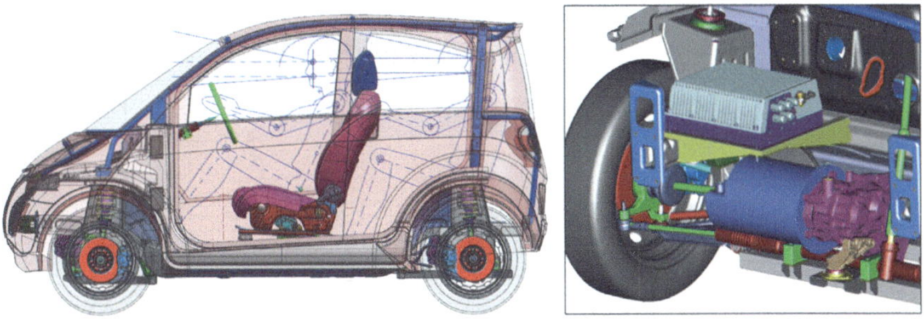

Fig. 10 WIDE-MOB/P-MOB vehicle (*left*) and HiWi motor integrated into vehicle (*right*)

5.1 Exchange Coupled Nano-Composite Magnets

Refinements to the nanostructure of NdFeB magnets and the substitution of some lattice atoms by other elements can have a dramatic impact on the magnetic properties of a material. However, pushing far beyond the theoretical limit to NdFeB energy product of 485 kJ m^{-3} requires something more. One route is to look towards composite materials which consist of phases each contributing their own desirable properties [5].

Ordinarily, the presence of soft magnetic phases in a permanent magnet will be heavily detrimental to its coercivity. Kneller and Hawig [6] proposed a new approach to fabrication combining a hard phase such as NdFeB, characterised by a high coercivity, and a soft phase such as α-Fe with a low coercivity but high saturisation magnetisation, into a bulk nanocomposite structure exhibiting both high coercivity and saturisation magnetisation.

The principle was derived from a simple one-dimensional model with alternating layers of hard and soft phases shown in Fig. 11.

The limit for this soft phase thickness was calculated to be 10 nm for an NdFeB/α-Fe composite. When this condition is met the hysteresis loop loses its characteristic step for a hard soft composite and the phases are said to be fully exchange coupled. Such composite structures are produced by rapidly quenching an alloy which is lean in Nd and rich in Fe such that the excess Fe forms into crystalline inclusions of α-Fe in an NdFeB matrix. The rapid cooling needed to produce the nanoscale microstructure required, means that conventional processing techniques Work in the HiWi project has concentrated on addressing this via the rapid annealing of 30 μm thick iron-rich amorphous NdFeB precursors with a view to building up successive layers additively. In order to develop this process route, melt spun ribbons of amorphous NdFeB material has been differing degrees of heat treatment in order to generate a nano crystalline microstructure.

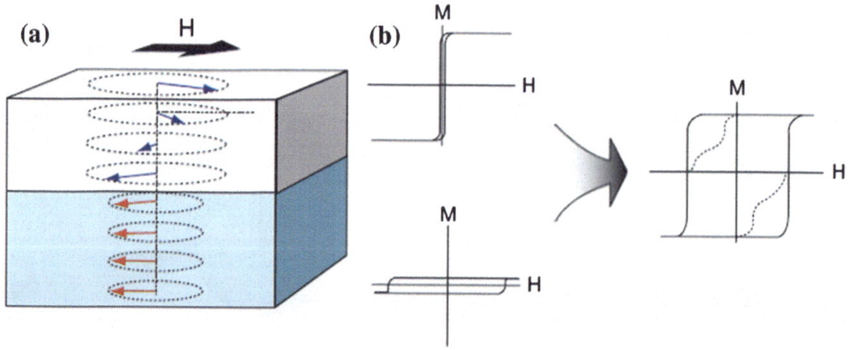

Fig. 11 **a** A bilayer structure with a hard phase bottom layer and a soft phase top layer. **b** Exchange coupling between a soft ferro-magnet with a tall hysteresis loop and a hard ferro-magnet with a wide hysteresis loop to form a high and wide hysteresis loop. The *dashed line* with kinks shows a decoupled composite in which the two loops are superposed [5]

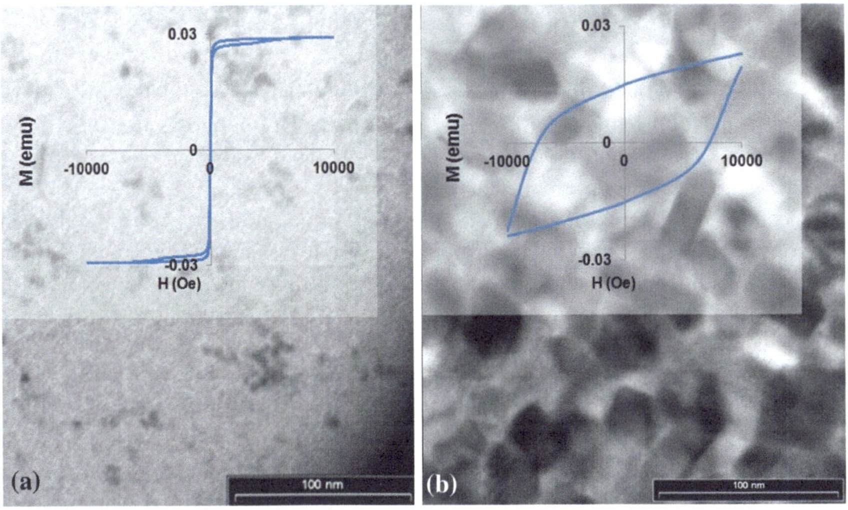

Fig. 12 TEM images with overlaid magnetic hysteresis *curves* showing **a** an undertreated sample and, **b** a fully annealed NdFeB magnet

Figure 12 shows images and the magnetic behaviour of two 30 μm thick samples of NdFeB material which have been differing degrees of heat treatment via laser annealing. The initial amorphous material behaves as a soft magnet with hysteresis increasing as crystallization takes place. As can be seen in Fig. 13b a nano scale micro structure has been successfully generated.

Fig. 13 SEM image showing a Dy deposit on a NdFeB magnet

5.2 Dysprosium Addition

It is established that additions of heavy rare earths (HRE), Dy and Tb enhance the coercivity of NdFeB magnets because the HRE-Fe–B compounds have a higher magnetic anisotropy than does NdFeB. However these additions also result in a considerable reduction in remanence.

Since magnetic reversal originates primarily from grain boundaries, targeted additions of Dy or Tb to the region close to grain boundaries will enhance the coercivity without reducing the remanence, resulting potentially in improved energy products and a significant reduction in HRE consumption (Particularly important for Dy given its cost relative to Nd).

One approach is to coat magnets with Dy by sputtering and then heat treat to allow Dy to diffuse into the grain boundaries of the substrate. This produced improved magnetic properties but was not suitable for mass production because of slow deposition times [7]. This has lead to a search for a technique for rapidly depositing a −5 μm layer of Dy onto NdFeB magnets in an industrial scale process.

Dy free NdFeB magnets have been coated with varying thicknesses of gas atomised Dy powder using a spraying process. Since the powder particles are larger than 5 μm in diameter, the desired volume of added Dy is achieved with a discontinuous coating as shown in Fig. 13. Interferometer measurements show that the sample shown has a deposited volume equivalent to a 5 μm continuous coating. This represents a significant reduction in Dy content over conventional alloying.

6 Conclusions

The substitution of driving cycles to key representative points in the torque, speed envelope and the development of new optimization tools has allowed drivetrain design to be optimised over a realistic usage case and has allowed the effect of a dynamic torque split across motors of differing characteristics to be assessed.

In the case of a light weight three person vehicle (the WIDE-MOB car) the most suitable topologies have been found to be a stator mounted permanent magnet machine which provides most torque at low speeds and a permanent magnet assisted synchronous reluctance machine which is more efficient at higher speeds.

The use of a dynamic torque spilt for the front and rear motors have been predicted to offer an improvement in efficiency of 0.95 % over the NEDC cycle.

The optimized motor designs have been manufactured and will be installed in a test vehicle.

Two new process routes have been investigated for the production of magnetic material with reduced rare earth content. The most promising route for the production of bulk magnets for use in electric drives is the deposition of Dy followed by annealing to induce diffusion of Dy down grain boundaries.

References

1. U.S. Geological Survey (2010) Mineral commodity summaries 2010: U.S. Geological Survey, 2010
2. Michel A (2004) The ARTEMIS European driving cycles for measuring car pollutant emissions. Sci Total Environ 334–335:73–84
3. Yuan X, Wang J, Colombage K (2012) Torque distribution strategy for front and rear wheel driven electric vehicle. In: Proceedings of the PEMD'2012 conference
4. Lazari P, Wang J, Chen L (2012) A computationally efficient design technique for electric vehicle traction machines. In: Proceedings of the ICEM'2012, Marseille, France
5. Liu J (2009) Nanoscale magnetic materials and applications. Springer, Boston
6. Kneller E, Hawig R (1991) The exchange-spring magnet: a new material principle for permanent magnets. IEEE Trans Magn 27(4):3588–3560
7. Hirota K, Nakamura H, Minowa T, Honshima M (2006) Coercivity enhancement by the grain boundary diffusion process to Nd–Fe–B sintered magnets. IEEE Trans Magn 42(10):2909–2911

eFuture—Safe and Efficient Electrical Vehicle

Frédéric Holzmann, Volker Scheuch and Pascal Dégardins

Abstract This paper presents the eFuture project that has been funded by the European Commission within the Framework 7—Green Cars Initiative. After a short presentation of the project and its objectives, the new vehicle concept will be presented and the physical integration as well. Later each function in the vehicle will be enlightened with its theory part and the corresponding results. Beside the virtual co-pilot and the driver coaching, the new electric driveline will be presented. The core idea of using a decision unit able to switch between driver and co-pilot, between energy saving and safety, and between most adequate actuators will be presented as well. Finally general results will be presented to light out the confirmation/ information of the hypothesis for our new vehicle concept.

Keywords European project · EV · Green ADAS · Torque vectoring · Decision unit

1 Presentation of the eFuture Project

The eFuture project [1] participates in the Green Cars Initiative in Europe. International connected bodies will focus from 2010 to 2013 on the development and validation of an adequate E/E architecture for safe and efficient vehicles. The partners are Intedis, Tata Motors European Technical Centre, Hella, Miljøbil, WIVW and IFSTTAR.

F. Holzmann (✉) · V. Scheuch · P. Dégardins
Intedis GmbH & Co KG, Max-Mengeringhausen-Straße 5, 97084 Würzburg, Germany
e-mail: frederic.holzmann@intedis.com

V. Scheuch
e-mail: volker.scheuch@intedis.com

P. Dégardins
e-mail: pascal.degardins@intedis.com

B. Müller and G. Meyer (eds.), *Electric Vehicle Systems Architecture and Standardization Needs*, Lecture Notes in Mobility,
DOI 10.1007/978-3-319-13656-1_6

The main question will be how to optimize the cooperation between an electric driveline and green ADAS for an optimal dynamical energy management. Main objectives of the project are:

- Development of an electric driveline with adaptive control
- Greening the ADAS
- Supporting the driver dynamically for optimal energy management
- Developing a lean and interconnected architecture for an electric vehicle

2 Vehicle Concept

Our project is in the continuity of former European projects like SPARC or HAVEit [2]. In those projects we had already split the vehicle architecture in a command layer and an execution layer with a single motion vector as interface. This concept came out of the integration of a drive-by-wire platform and a virtual co-pilot in the first project. Here a first decision unit has been designed centrally to switch binary between driver and virtual co-pilot. Later in the second project the decision unit has been extent to deal with gradual automation level. Here again the decision unit will be extended by adding a new dimension: the efficiency of the vehicle as it can collide with the safety must for the vehicle. Therefore the vehicle shall have the possibility to select gradually on one hand the split between driver and co-pilot and on the other hand the split between efficiency and safety. With this solution it would be better to extend the EV range by use of efficient driveline and efficient ADAS without endangering the road users.

2.1 Functional Architecture

The functional architecture follows the known separation into command and execution layer, supplemented by a perception and an energy layer. The four layers, or domains, are attributed to dedicated tasks in the architecture of the vehicle:

- *Perception Layer*—Collection of all environmental vehicle data via exterioceptive sensors (cameras, radar, navigation, eHorizon) and processing the data for use in the command layer. The driver is assigned to this layer as he performs the same action like the sensors plus data fusion: perceiving and interpreting outside signals.
- *Command Layer*—Selection of most suitable trajectory for the driver's trip. There are two different ways to create commands on where to drive the vehicle: either manually by the driver or semi-automated by ADAS functions. The task of balancing between both is performed by the Decision Unit 1.

- *Execution Layer*—Selection of the most suitable actuators for executing the trajectory command. By comparing the command with the current vehicle status (stability, battery) the Decision Unit 2 forwards the appropriate signals to the drivetrain and chassis actuators.
- *Energy Layer*—Surveillance and optimisation of the energy flows. The battery status and the activated consumers define the strategy of enabling and shaping energy flows by setting current and acceleration limits to the Command and Execution layer and switching on/off individual consumers.

2.2 E/E Architecture and Vehicle Head Unit ECU

The E/E architecture has been implemented as extension of the current Tata Vista vehicle. Thus we have on one side the components that are corresponding to the host vehicle and the new components. The interconnection of these two different worlds is done via a gateway, which is integrated into the Vehicle Head Unit (VHU). This ECU corresponds at the same time to the main control for the hosting of the new functionalities (e.g. ADAS, decision units). Behind that new main component, dedicated CAN networks have been rolled out for each specific domain: ADAS, electric driveline, HMI etc. With this approach it is possible to increase the perimeter over the project time by using discrete components.

The VHU is the central control unit for the electric vehicle. At first glance it controls all extra equipment of the EV compared to a common car with a combustion engine. Furthermore it is the platform for the advanced functions. The VHU basically is a multi-controller board with four independent 32-bit microcontroller systems. Each system has its own oscillator circuit, and the systems use an asynchronous communication CAN bus, so they can't block each other. A multiprocessor concept with four microcontrollers of the same type is the basic platform for all software functions of the VHU. In addition, an AUTOSAR compatible system allows re-use of existing software modules: e.g. hardware drivers and the operating system. Model based software development is made easy, the interfaces for application software is already defined. Four complete autarkic μC-systems connected with an private asynchronous data bus on the one hand gives every software developer the whole microcontroller performance, and, on the other hand reduces the risk of mutual interference.

Following the automotive standard, both, the Accelerator Pedal as well as the Shifter, are equipped with two redundant potentiometers. A single failure in any sensor or in any signal-processing path or in any connected μC can be tolerated without the risk of losing the sensor information.

In addition to their dedicated functions, μC work together to improve the robustness of the drive train functionality. Because of mutual monitoring, both microcontrollers are informed about the proper function of each other. In case of a

Fig. 1 VHU with four Bolero μCs, shown with housing and cover

faulty μC (e.g. missing CAN signals) the other microcontroller can override it and switch the drive motors into a limp home function. For that reason, they both need all necessary input information, of course buffered by preamplifiers, so they can't crosswise block the input signals (Fig. 1).

3 Command Layer

In this section the command layer will be presented. It corresponds to the virtual co-pilot (ADAS modules) and to the interconnection with the real driver—the HMI. It has its own decision unit for switching between energy and safety and between driver and virtual co-pilot. Our upfront perception layer will not be presented as it is based on state of the art technology.

3.1 Virtual Co-pilot (ADAS)

The virtual co-pilot aims at defining the motion vector, in term of speed objectives, longitudinal acceleration and yaw moment to be applied. The decision to apply or not apply to the request motion vector is then done by the Decision Unit. This motion vector, depending on the available information, must ensure the safety of the vehicle and optimise the consumption, with the constraint of an embedded platform for the execution. In order to simplify the problem, we have decomposed the problems under longitudinal and lateral aspect. Lateral Advanced Driving Assistance System (ADAS) helps the driver by warning him, or maintaining him on the lane. It requires, from the perception level, to have the vision of the road from the camera system. The longitudinal ADAS are more depending on the sensor availability and on the selected mode.

ACC corresponds to Adaptive Cruise Control and is defined by ISO norm 15622. It relies on the perception of a front obstacle and on the knowledge of the ego vehicle dynamic. The perception of the front obstacle is an output of the perception system. It results from the fusion of the Radar perception and of the stereo camera perception. In addition the common ACC is modified to optimise the consumption in respect of the direct environment: SAGA is Smart And Green ACC. This last function takes into account both: near environment (from radar and camera) and far environment (from eHorizon) to optimise the vehicle speed. As they are derived from ACC, they must handle the two operating modes of ACC, namely distance following and speed following.

3.1.1 Speed Control

On speed control, the optimised speed uses the eHorizon to know the future of the road, as geometric description, legal speed limit and also to integrate several rules accordingly with the road topology. From this information several optimisation processes were investigated to define a consumption optimised speed profile, which takes also into account the regeneration capacity. The function regulates the speed around the drivers desired speed. If the eHorizon is available, the drivers desired speed may be higher than the optimal speed. In this case, the optimal speed is chosen.

The Fig. 2 shows the resulting speed profile, from ACC and SAGA, without interaction with a front vehicle. We can see that the 1 km trip needs 48.9 s for the ACC and 55.7 s for the SAGA—this is a 7 s difference. For the SoC, the first speed profile will consume around 0.1 kW, while the SAGA can regenerate almost all energies during numerous deceleration requests.

3.1.2 Distance Control

How should electric braking to a target speed be done? Basically, two different strategies are possible: (a) 'recuperation before sailing' with the vehicle braking

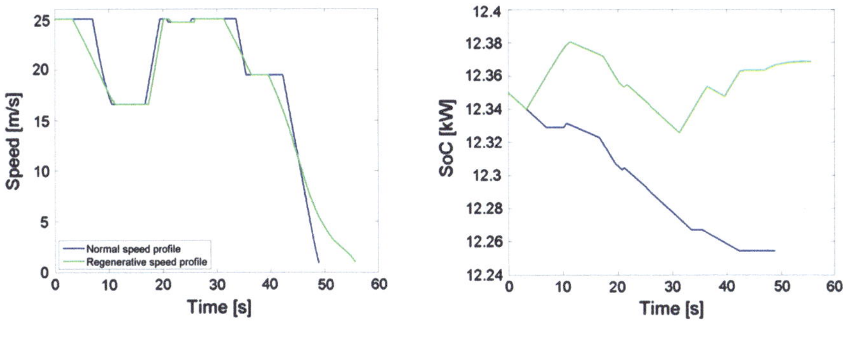

Fig. 2 Speed profile and resulting evolution of the state of charge (SoC)

electrically and a minor adjustment to target speed afterwards and (b) 'sailing before recuperating' with the vehicle sailing up to the point where maximum recuperation is necessary in order to reach target speed in due time. Testing these two strategies in the driving simulator yielded the result that recuperation before sailing is more energy efficient than the other strategy. An example for decelerating the vehicle from 70 to 30 km/h within a road section of 200 m brings a recuperated energy of 40.1 Wh for recuperation-sailing compared to just 35.7 Wh for sailing-recuperation. Therefore, this characteristic of the vehicle is implemented in the parameterization of the ECC system (Fig. 3).

We have evaluated different braking strategies with the safety as a first requirement. In order to evaluate safety, we compare the safety domain, meaning the domain of speed and distance that allows to use the function and to brake with the considered deceleration while approaching a slower vehicle. In Fig. 4, we define the safety domain for a front vehicle that drives at 20 m/s and a desired headway of 2 s.

We can clearly see that the domain for the regenerative braking ACC is smaller than the conventional ACC. Using the regenerative deceleration only, it is not possible to obtain the same safety level as a conventional ACC with the same use case (initial time headway of 2 s, speed of 30 m/s and final speed of the lead vehicle of 7 m/s).

The possible solutions for implementation are:

- To increase the minimal time headway that is defined by the driver. To obtain the same safety, we need to increase the minimal time headway up to 5 s. The resulting distance is hardly achievable as the sensor range is limited and other road user may cut in the space between vehicles.
- To switch to conventional braking if the time headway drops below a given threshold. For instance, we can maintain a collision free system, as for a conventional ACC, with an initial time headway of 3.7 s and an activation of a stronger braking at a threshold on the time headway of 1.5 s.
- To switch to an emergency braking if the deceleration of the lead vehicle and the distance drops below given thresholds.

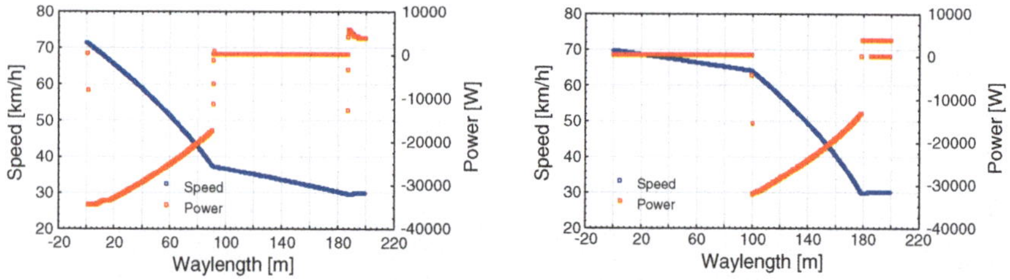

Fig. 3 Speed profile and recuperated power for the two different strategies that can be used to reach a new target speed (*left* 'recuperation before sailing'; *right* 'sailing before recuperation')

Fig. 4 Conventional ACC and regenerative braking ACC safety domain

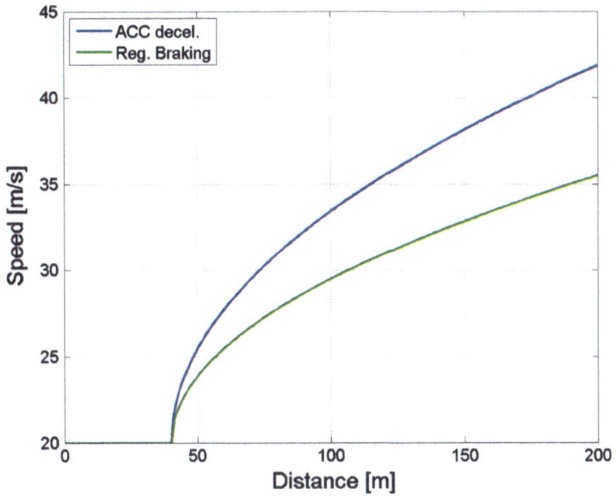

We have implemented the second option and achieve very good regeneration while maintaining a good safety. For instance, if we consider the scenario where we approach a slow vehicle, we have a difference of 0.12 kWh on the State of Charge of the battery (see Fig. 5).

3.2 *Human Interface for Coaching*

The driver coaching comprises coaching advices for the following situations: speeding, hard acceleration, car following with too small headway, approaching a lower speed limit, passing through a curve, approaching and passing a hilltop, downhill driving. Inefficient driving behaviour (e.g. frequent hydraulic braking) is not part of the coaching function as only aspects of the driving behaviour are included that are not safety relevant or may not lead to safety critical situations.

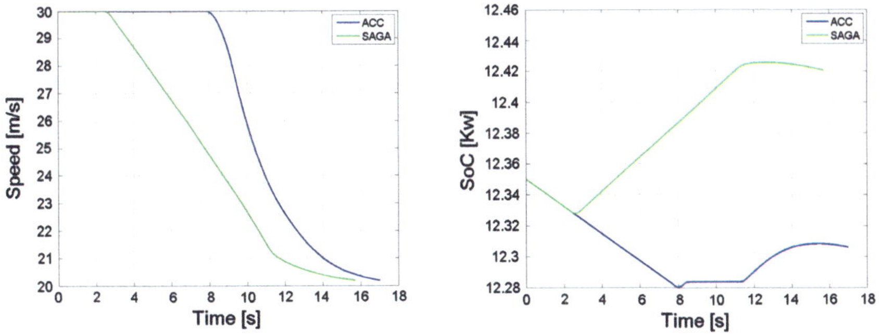

Fig. 5 Approaching a vehicle comparison of conventional ACC and regenerative braking ACC

A driver coaching study has been conducted which evaluates the driver coaching concept on the one hand and compares the driver coaching concept with two further coaching conditions on the other hand: a short briefing on efficient driving prior to the drive or an unspecific feedback during the whole drive (current consumption relative to an efficient reference). Finally, the driver coaching study was conducted in order to address the following research questions:

- Which impact has specific online coaching on energy consumption/energy efficiency?
- How much do drivers benefit from specific online coaching compared to unspecific online coaching or to verbal-only instruction?
- How far does the specific online coaching change the driving behaviour?
- How do drivers evaluate specific online coaching? How strenuous is specific online coaching? How usable is specific online coaching?

By means of the eFuture driver coaching function the energy efficiency was significantly increased. Compared with the baseline drive, participants improved their driving style significantly. However the drivers in the verbal instruction group did not improve their driving skills as much as the others. Drivers in the coaching condition consumed 7.2 % less than other drivers in the verbal condition. The consumption scale improved the energy efficiency in a comparable way as the specific online coaching. Though both groups benefit in a comparable way from online feedback, interesting results are provided when looking at the driving parameters. Drivers in the coaching condition tend to drive slower in the second drive compared to other drivers and the sailing time is significantly lower.

When it comes to the workload aspect, coaching advices and the consumption scale score rather low. Neither feedback option is rated to be rather disturbing or distracting. However, though the consumption scale is viewed to be slightly more motivating than coaching advices, it is also rated to be slightly more disturbing. Regarding the acceptance both, the coaching advices and the consumption scale, were highly accepted. When using an unspecific feedback (as the consumption scale), however, it is advisable adding some more information about the savings in form of saved kWh, gained kilometres or in form of saved money. This was mentioned and wished from several participants.

The idea behind the recuperation advice was rated very positive. However, participants insisted that this recuperation advice has to appear very precisely. This is an important factor as on the one hand, the more precise the advices are presented the more energy can be saved, but on the other hand the acceptance is increased which in turn increases frequent usage of the driver coaching function. Most important improvements therefore must be made regarding the correctness and precision of the respective advices.

Though unspecific feedback provides comparable energy savings as specific online feedback, in summary, the presented results lead us to the conclusion that we prefer and recommend the specific online coaching by means of coaching advices over unspecific feedback. This is mainly due to three reasons e.g. drivers were distracted by the unspecific feedback. In times of increasing distraction of the driver by means of various assistance system or mobile devices, further distraction must be kept as low as possible. Therefore, the unspecific feedback has a clear drawback. Further, coaching advices guide drivers towards a variety of driving behaviour patterns, whereas the unspecific feedback mostly resulted in lower velocities and sailing. However, energy savings can be accomplished by more than lower velocities and sailing. Those driving behaviours have a high potential to increase energy efficiency significantly when applied correctly. And last, the concept of the eFuture driver coaching function comprises one important point, namely to give drivers the opportunity to drive advice free. This is not possible with the unspecific feedback.

4 Execution Layer

The execution layer is focusing on the vehicle movement by means of the algorithms to estimate and control the vehicle movement and as well all the actuators.

4.1 Decision Unit

The main task of DU2 is to distribute the actuator requests and to care for the vehicle safety—related to actuators. In the eFuture concept the DU2 distributes the vehicle commands between the steering, the hydraulic brakes and the electric motors. Regarding safety concerns of the real prototype the steering and hydraulic brakes are not influenced by the eFuture software in order to have a well-known and safe emergency level for the driver. So in the real prototype the DU2 cares about the electric motors. The main tasks are to limit the power consumption of the electric motors, if this is requested by the vehicle Energy Management. Together with the Power Modes function the DU2 cares about the activation of the electric motors and reacts to motor failures and communication disruptions. DU2 processes failures reported by the electric motors and switches to different limp home modes. If one or both motors report a warning, both motors are limited to 300 N m in order to preserve the electric drivetrain. If one motor reports a severe error, this motor is shut down and the other motor is limited. So the driver is able to manoeuvre the vehicle to a safe place.

4.2 Torque Vectoring

Torque vectoring processes and distributes motion vector request as torque commands to the wheels of the vehicle. Besides the vehicle dynamics the controller has to consider the limited wheel forces and wheels independencies. For the controller design a linear parameter-varying (LPV) control strategy is used. The controller deals with the nonlinearity of a vehicle and it is possible to tune the controller for different operation points like low velocity, high velocity, straight line driving, strong cornering, and so on. Additional an anti-windup scheme is used to deal with the physical limitations of the electric motors and is later extended to a torque and wheel slip limiter (TSL) to cope with spinning and blocking wheels.

To evaluate the vehicle performance several tests have been performed. In order to show the influence of Torque Vectoring a constant radius turn is helpful. In Fig. 6 the tyre slip angle "alpha" for a constant radius turn is shown. During this test, the driver steers the vehicle on a circle with 15 m radius and increases the velocity slowly. The main difference between the vehicle with equal torque (eqTor) and Torque Vectoring (TorVec) is visible between 15 and 20 s. For the eqTor, the difference between the front tyre slip angle "alphaFront" and the rear tyre slip angle "alphaRear" increases up to 0.2 rad (at 19 s) whereas the difference between the front and rear tyre slips for TorVec stays below 0.06 rad.

Thus torque vectoring is useful to adapt and improve the vehicle characteristics during driving. Many parameters can be tuned but to find an optimal solution, some

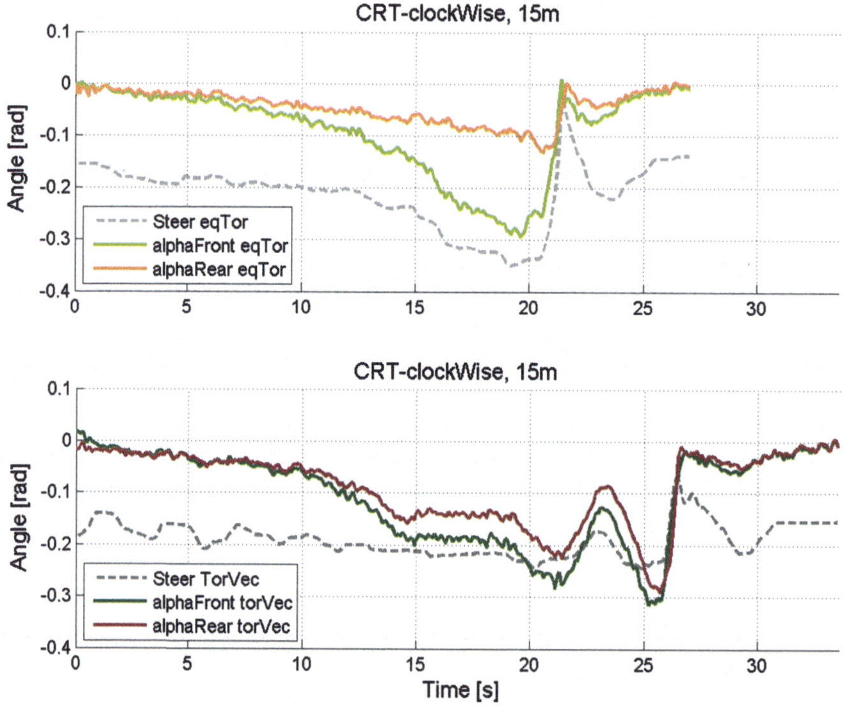

Fig. 6 Constant radius turning—tyre slip angle

effort has to be done. In the eFuture parameterisation the vehicle deals with electric motor saturations, tyre limitations and different lateral vehicle behaviour for different velocities.

4.3 Electric Drives

The motors are mechanically unconnected in our vehicle to improve vehicle control efficiency as presented in the last section. Hence special care has to be taken that unwanted differences in torque don't cause the car to swerve or skid.

To increase the robustness and safety of the drive train system both inverters talk to each other and both stop applying any torque in case of a problem appearing in one of them. This is realised by exchanging a 1.2 kHz square wave signal during normal operation mode. In case an inverter detects that the signal of the other inverter is missing, it stops its signal transmission and initiates the "safe state": no torque is applied. One main requirement to the inverters is, that the detection of a wrong or missing wave signal, the interruption of the own sending of the wave signal and the transition into the safe state has to be achieved within a very short time frame. This point-to-point exchange of square wave signals is additionally monitored by μC-3 and μC-4 in the VHU (Fig. 7).

In case of an erroneous torque reduction after an inverter having problems the VHU can find out which inverter is the reason for the fault. It is a conceptual approach to override the missing square wave signal with optional VHU drivers and applying an artificial signal to the still operational inverter to enable a limp home mode. The VHU's square wave drivers can also force the inverters into no torque mode, if the motors behave different as commanded via CAN.

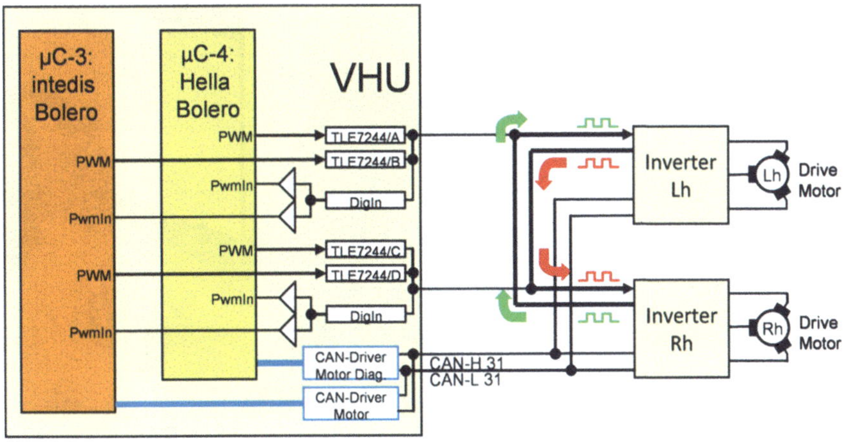

Fig. 7 μC-3 and -4 redundant interfaces to drivetrain

4.4 High-Voltage Battery and Battery Management

A system with 2 parallel connected battery strings with individual current measurement systems will be used. The high-voltage current sensor communicates over the LIN bus with the BMS. Each of the battery strings are built up of 6 identical battery modules connected in series. The total amount of cells in each module is 30 (3P10S, i.e. 3 in parallel 10 in series). The battery configuration is to use 1 front row (FR) and 1 rear row (RR) in parallel (360 cells) (Figs. 8 and 9).

The main characteristics of our high-voltage battery are compiled below:

- Maximum energy available at 20 °C: 26 kWh
- Battery pack voltage: 180–222–247 V
- Maximum current draw: 200 A
- Weight <250 kg
- Peak Power (100 % SoC) at 20 °C: 47 kW
- DC instantaneous (10 s) impedance at 25 °C: <55 mΩ

Fig. 8 Main battery building block

Fig. 9 Complete battery tray without lid. The cables on top are the two strings of the daisy chain harness connecting the cell monitors and the main controller, which resides in the front middle

5 Main Results

After a debugging phase, the two vehicles have been intensively used for validation purposes on different testing facilities. A system FMEA accompanied the final integration. The results obtained according to the validation plan prove the achievements for safety and efficiency in the eFuture electric vehicle:

- Successful implementation of a robust functional architecture with a consistent function hierarchy and signal logic together with a concept of Decision Unit serving as the central control functions
- Creation and implementation of a comprehensive Functional Safety Concept for the functional architecture
- Introduction of functions for an efficient and safe operation: Smart ACC assists the drive to maintain a safe speed, Torque Vectoring eases dynamical situations, and the Decision Unit adapts the drivetrain to different driving situations
- Favourable integration of an Energy Management for supporting smart and economical current draw
- The integration of the driver into the vehicle's decision loop is the most promising element for enhancing efficiency. The precise observation and careful coaching of the driver leads to an enormous increase of accepted assistance and thus, efficiency.

The achievements of eFuture clearly show that a new EV architecture enables a high safety level and promises ways to less energy consumption. We believe that the combination of different measures is the key for achieving a satisfying range extension of the electrical vehicle.

This paper is a summary of all public deliverables that have been as well written by/with the partners. We would like to thank them for the fruitful cooperation despite we cannot name everybody here.

References

1. http://www.efuture-eu.org/
2. http://www.haveit-eu.org

HEMIS Project (Electrical Powertrain HEalth Monitoring for Increased Safety of FEVs): Limitations of Electromagnetic Standards for Vehicles Equipped with Electrical Powertrain

Alastair R. Ruddle, Rob Armstrong and Ainhoa Galarza

Abstract The HEMIS project is concerned with health monitoring of electrical powertrain in order to enhance the maintainability and safety of electric vehicles. Electromagnetic emissions from electrical powertrain will change the electromagnetic environment in vehicles, which may have implications for electromagnetic compatibility and, because of the high electrical power levels, human exposure to electromagnetic fields. This paper outlines some of the electromagnetic issues under investigation in the project, including limitations of existing test methods for electromagnetic compatibility, and a lack of test methods for the assessment of in-vehicle human exposure, as well as for both electromagnetic compatibility and human field exposure issues relating to wireless charging of traction batteries.

Keywords Electrical powertrain · Electromagnetic compatibility · Human electromagnetic field exposure · Standardization

1 Introduction

In order to achieve the 2020 target of reducing energy consumption and CO_2 emissions, electric vehicles need to reach a significant market share. However, this objective presents manufacturers with new challenges due to the scarcity of

A.R. Ruddle (✉)
MIRA Limited, Watling Street, Nuneaton, UK
e-mail: alastair.ruddle@mira.co.uk

R. Armstrong
York EMC Services Limited, University of York, Heslington, York, UK
e-mail: rob.armstrong@yorkemc.co.uk

A. Galarza
CEIT-IK4, Paseo Manuel de Lardizabal, 15, San Sebastian, Spain
e-mail: agalarza@ceit.es

B. Müller and G. Meyer (eds.), *Electric Vehicle Systems Architecture and Standardization Needs*, Lecture Notes in Mobility,
DOI 10.1007/978-3-319-13656-1_7

reliability data for the new technologies involved, the most notable of these being the electric powertrain.

Although electric motors have been used in industry for a long time, and land speed records were broken by electric vehicles at the end of the nineteenth century, the use of electric motors for passenger vehicles has only recently experienced a revival of interest. Modern electric vehicles need to take advantage of light-weight materials and more efficient powertrains in order to achieve the level of autonomy that drivers now require. The relative immaturity of the technologies involved means, for example, that the lifespan and potential failure mechanisms for the components are not yet fully understood. According to opinion polls [1], concern about the reliability of fully electric vehicle (FEV) technology is one of the top three reasons why 58 % of potential consumers would choose to use a hybrid vehicle rather than an FEV.

When the time comes to study the safety of electric vehicles, electromagnetic compatibility (EMC) issues must be taken into account, since electric powertrain components are additional sources of electromagnetic fields. Furthermore, the possible impact of the powertrain's electromagnetic emissions on the occupants should also be considered. However, current automotive EMC standards originated when vehicles systems were predominantly mechanical. Thus, this paper outlines a review of current automotive standards for vehicles with electric powertrains, regarding both EMC and human field exposure issues. Exposure during wireless charging of traction batteries is also considered.

1.1 HEMIS Project

The main objective of the HEMIS project is to enhance the maintainability and safety of full electric vehicles, by developing an in-vehicle Prognostic Health Monitoring System (PHMS) for the electrical powertrain (see Fig. 1). The PHMS will be able to analyze the failsafe state of the powertrain. It will also perform failure diagnosis and estimate the remaining useful life of the powertrain, including which component will fail first, thus enabling a predictive maintenance policy.

The proposed assessment of the powertrain condition is based on in-vehicle sensing of suitable physical parameters. In order to select these parameters, firstly the most critical failure modes of the powertrain components have been identified [2]. Secondly, the suitability of various physical characteristics for failure mode detection, diagnosis and prediction has been assessed. Currently, experimental tests are being performed in order to collect datasets relating to the evolution of these variables evolution during the relevant degradation processes.

For this application, it is considered that changes in magnetic flux or electric field in the vicinity of the motor or control system could be indicative of a degraded operating state, which may lead to premature failure of a drive system component or to electromagnetic interference with the electrical systems of the vehicle. Hence, the PHMS will also monitor the in-vehicle electromagnetic emissions. In addition, the

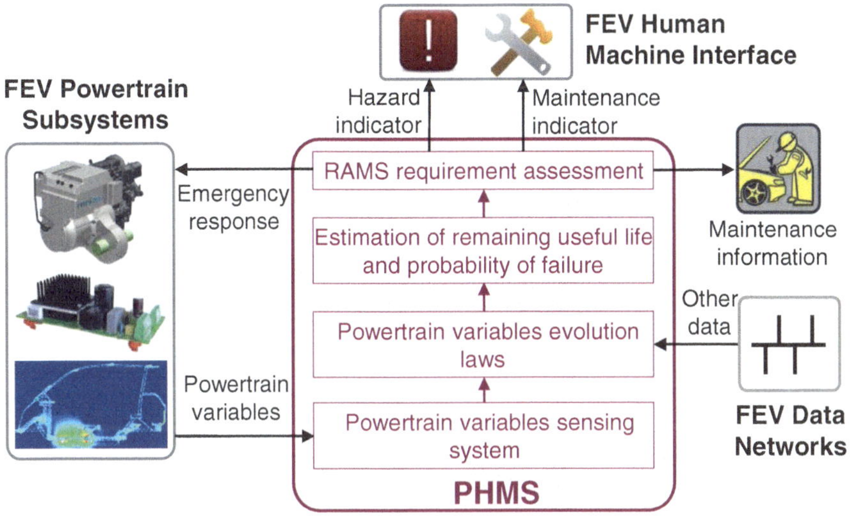

Fig. 1 The HEMIS PHMS concept

HEMIS project will also take into account the possible effects on the human body, and will assess whether these fields, which may differ from those experienced in conventional cars, could exceed exposure limits. In that case, suitable reference levels will be proposed and the PHMS will also monitor that the electromagnetic field in order to ensure that occupant exposure remains below acceptable limits.

2 Electric Vehicle EMC Testing

Automotive EMC (electromagnetic compatibility) standards originate from a time when the main threat posed by radio-frequency noise from vehicles (arising from spark ignition) was interference to broadcast transmissions, while the vehicle systems were predominantly mechanical with few potential immunity concerns. Since that time, on-going technological developments have resulted in changes in the nature of both the emissions that vehicles generate and the radio-based services under potential threat, as well as steadily rising deployment of electronic sensors, actuators and control systems that must continue to function in an increasingly complex electromagnetic environment. Rapid changes in the technologies associated with the electrification of vehicle powertrain, in particular, raise questions as to whether the existing test methods remain appropriate for such vehicles. This section therefore outlines a review of current automotive EMC standards, as well as anticipated future developments, in relation to vehicles with electric powertrain. A more detailed description of this analysis is given in [3].

For the purposes of type approval, requirements defined in the Automotive EMC Directive (2004/104/EC [4]) and UNECE Regulation 10 [5] could be used.

This creates some anomalies as these documents reference different versions of other standards. However, 2004/104/EC will be replaced by UNECE Regulation 10 with effect from 1st November 2014 [6]. Consequently, this review focuses on UNECE Regulation 10 as the main source of automotive EMC test requirements. Both 2004/104/EC and UNECE Regulation 10 include emissions test modes for vehicles with electric powertrain. Recent amendments to UNECE Regulation 10 also include requirements for testing rechargeable energy storage systems (RESS).

2.1 Component Level EMC Tests

In the automotive industry EMC testing is carried out on vehicle sub-systems (described as *component level*), as well as at whole vehicle level. At component level, the test requirements include evaluation of emissions and immunity phenomena due to both radiated and conducted coupling. The test methods developed for this purpose assume that the sub-system is small and will be mounted in a metallic vehicle. However, the RESS and electric powertrain packages may be physically large, electric vehicles often have non-metallic body panels, and the high-voltage power networks linking the RESS and electrical powertrain components have very different characteristics to conventional vehicle wiring.

Component immunity is covered by the ISO 11452-X series of standards [7], which describe a number of test methods covering specific frequency ranges, any of which may be used to demonstrate compliance provided that the full frequency band (20–2,000 MHz) is covered. A number of limitations of the component level immunity test requirements (some of which are also relevant to conventional vehicles, as well as those with electrical powertrain) are summarized in Table 1.

For emissions at component level UNECE Regulation 10 references the corrected 2nd Edition of CISPR 25 [8], although there is a more recent 3rd Edition. UNECE Regulation 10 states that the sub-system should be in *normal operating mode, preferably maximum load*. However, maximum load may not necessarily generate the highest emissions. It is noted in [9] that the load has a significant

Table 1 Gap analysis for component level immunity testing

Gaps	Future standards
Choice of test procedures could lead to selection of 'easier' tests with lower limits	Not addressed
Test methods in different parts of ISO 11452 are not equivalent	Not addressed
Use of vertical polarization illumination only	Not addressed
Size of the ground plane used (FEV RESS and electric powertrain packages are potentially very large)	Not addressed
Ground plane materials (not necessarily representative of composite body-shell materials)	Not addressed
Cable harness arrangement for traction package	Not addressed

Table 2 Gap analysis for component level emissions testing

Gaps	Future standards
Load influence on emission levels	Amendments to CISPR 25 (draft 4th edition)
Instrumentation tailored to combustion engines	Not addressed
Load simulator not defined for traction package	Not addressed
Artificial Network definition not applicable to high voltage networks	Not addressed
No radiated emission limits below 30 MHz	Not addressed
Size of the ground plane used (FEV RESS and powertrain packages are potentially very large)	Not addressed
Ground plane materials (not necessarily representative of composite body-shell materials)	Not addressed
Cable harness arrangement for traction package	Not addressed

influence on the emission levels for conducted and radiated emission measurements on electric vehicle propulsion systems.

Thus, it is expected that there will be changes in the 4th Edition of CISPR 25 that take account of the features of electric powertrain. A summary of limitations of component level emissions requirements with respect to electrical powertrain sub-systems is given in Table 2.

2.2 *Vehicle Level EMC Tests*

Vehicles are tested for immunity to radiated electromagnetic fields in the frequency range 20–2,000 MHz [5], and are expected to meet the functional failure criteria defined for the various immunity related functions. A minimum set of criteria are specified in [5], but other vehicle systems that could affect the immunity-related functions must be tested using a method that is to be agreed between the manufacturer and the technical service.

The vehicle immunity test requirements (which are based on the ISO 11451-X series [10]) are generally just as applicable to vehicles with electric powertrain as to conventional vehicles. However, the pulse criteria for conducted disturbances that are listed in UNECE Regulation 10 only relate to 12 and 24 V systems, while electric vehicles may operate at up to 650 V [11]. So far, the only difference introduced in UNECE Regulation 10 in relation to the immunity of electric vehicles is an additional test mode in which the RESS is charged from an external supply via a cable. Wireless charging of RESS, however is not addressed.

Current vehicle-level radiated emissions test requirements [5] reference the amended 5th Edition of CISPR 12 [12], although there is a more recent 6th Edition. Broadband and narrowband emissions measurements are specified over the band 30–1,000 MHz, using an antenna at a fixed point on either side of the vehicle on a line through the engine (usually through the front axle). However, one of the features of vehicles with electric powertrain is that they offer much greater flexibility in terms of the location of electrical machines (which could be a single machine, one for each axle, or one for each wheel), their power electronics, and range extenders (which could include fuel cells, combustion engines or gas turbines). Thus, the sources of emissions may be much more widely distributed around the vehicle than in conventional vehicles, making the current test positions questionable. Other vehicle emissions test practices that are considered to be questionable include:

- no requirement to identify the "maximum" emission direction;
- use of a 20 dB conversion factor between peak and quasi-peak measurements, although the latter is actually dependent on the pulse repetition rate;
- use of a 10 dB conversion factor between measurements at 10 and 3 m distance, although this has been shown to be an over-simplification for extended sources such as vehicles (e.g. [13]).
- no requirement to test radiated emissions at frequencies below 30 MHz.

Broadband measurements on internal combustion engine vehicles are carried out with the engine running at constant speed (1,500 rpm), primarily to detect emissions from the spark ignition system. Vehicles with electric powertrain are to be operated at 40 km/h (or maximum speed if this is lower), either on a dynamometer under negligible load or in a free-wheeling mode with the driven wheels raised on non-conductive axle stands. An additional requirement for hybrid vehicles is that both mechanical and electrical drive systems should be operational during the measurements where possible. However, it is possible that the worst case scenario for emissions could be under high or partial load conditions. It is noted in [9] that further clarification of conducted and radiated emission measurements for vehicles with electric powertrain is necessary in the draft 7th Edition of CISPR 12, particularly as the load may have a significant influence.

A number of limitations of vehicle level emissions test requirements (some of which are also relevant to conventional vehicles, as well as those with electrical powertrain) are summarized in Table 3.

Table 3 Gap analysis for vehicle level emissions testing

Gaps	Future standards
Scanning for maximum emission	Not addressed
Receive antenna location	Amendments to CISPR 12 (draft 7th edition)
Questionable 20 dB correction factor for quasi-peak measurements	Not addressed, but test labs are avoiding use
Operating mode for electric vehicles	Not fully addressed—some requirements in later issues of CISPR 12
Load influence on emission levels	Amendments to CISPR 12 (draft 7th edition)
Questionable 10 dB correction factor for measurements at 3 m distance	Not addressed
Wireless battery charging	Revision of draft IEC 61980
Low frequency emissions (below 30 MHz)	SAE J551-5 [14] for 150 kHz–30 MHz

3 Electromagnetic Field Exposure in Electric Vehicles

The occupants of vehicles with electric powertrains are exposed to low frequency magnetic fields arising from currents flowing in the high voltage power network, traction batteries, and associated devices such as inverters and electrical machines. Power levels up to more than 200 kW are reported for hybrid/electric cars in [15]. In addition, there is growing interest in wireless inductive charging of vehicle traction batteries, which may be carried out while the vehicle is in use, posing field exposure risks for both vehicle occupants and bystanders. Thus, there is a need to consider possible health impacts of exposure to these fields, and to take account of these issues in the design of vehicles and wireless charging systems.

3.1 Field Exposure Limits

At present there are no product standards relating to in-vehicle field exposure, although work on a new standard for the measurement of low frequency magnetic field exposure in vehicles (IEC 62764) has recently been initiated. Nonetheless, there are general recommendations that have been published by various national and international bodies. In Europe, there is an EC recommendation (1999/519/EC [16]) for general public exposure that is based on limits [17] defined by the International Commission for Non-Ionizing Radiation Protection (ICNIRP) in order to protect against well-established acute physiological effects. Furthermore, an EC directive concerning occupational field exposure [18] is now due to be adopted in national legislation by July 2016. The low frequency limits of this directive are partly based on revised ICNIRP guidelines [19] for the band 1 Hz–100 kHz. In this frequency range the limits aim to avoid undesirable electro-stimulation effects. As this is an instantaneous phenomenon time-averaging is not considered to be appropriate [16, 19] and the worst-case exposure should be assessed.

The exposure limits of [16–19] are specified in terms of in-body quantities, which are not easy to determine. More readily measurable “field reference levels” have therefore been derived from these exposure limits. Thus, if the environment complies with the field reference levels then compliance with the exposure limits can be assumed. Consequently, magnetic field measurements have been carried out on a number of representative electric and hybrid vehicles (e.g. [11, 20]). A common feature of these measurements is that the highest fields tend to be found where the body is close to high power cables (often near the feet of the driver or front passenger) or traction batteries (e.g. around the lower back for rear passengers in vehicles with batteries located in the rear). Batteries are also mounted below the occupants in some vehicle architectures, while the electric motor is often located at the front axle, although multiple-motor architectures are also possible and in-wheel motors may become the standard approach in future.

Additive effects must be considered in human field exposure evaluations, with the result that the greater the number of frequencies that are present, the lower the levels that can be tolerated for any of them relative to the field reference levels. Furthermore, the influence of other fields in the environment also impacts on what can be tolerated from equipment generating fields.

High voltage power cables carry current transients with finite frequency components that need to be taken into account. Simply adding the field values relative to the reference levels, which must not exceed unity, would be excessively conservative as it assumes all components to the in phase. For non-sinusoidal exposures, therefore, [17] proposes a more realistic time-varying field exposure measure (the "weighted peak") that is based on an inverse Fourier transform. This not only takes account of the additive effects of multiple frequencies, but also reflects the relative phase of the spectral components. The use of this approach to assess measured and simulated in-vehicle magnetic field waveforms is described in [20].

The existing field reference levels have been derived for a standing human in an open environment under a uniform field. Consequently, the use of these reference levels in vehicles, where the occupants are seated, surrounded by nearby structures, and exposed to highly non-uniform fields, may be questionable. It is noted in [16, 17] that the field reference levels are intended to be spatially averaged over the body, provided that the basic restrictions are not exceeded, although neither provide any suggestions concerning acceptable levels of non-uniformity. Thus, the use of spatially averaged field values to describe non-uniform, non-localized exposures is anticipated in the exposure guidelines, but ICNIRP consider [19] that standards bodies should provide guidance on the use of spatial averaging and derive new reference levels for specific types of non-uniform exposure. The in-vehicle environment seems a prime candidate for this.

3.2 Active Implantable Medical Devices

Active implantable medical devices (AIMD) are devices that rely on a power source not provided by the body or gravity and are designed to be introduced into the body with the intention to remain there following the procedure. Examples include cardiac pacemakers, cochlear implants, stimulators, and infusion pumps.

It is noted in [16, 17] that compliance with field reference levels recommended for the general public cannot be assumed to ensure that AIMD will continue to function correctly. The EMC requirements for these devices are specified in the EN 45502 series of standards [21]. Electronic article surveillance systems and metal detectors (hand-held and walkthrough) are reported to have interrupted the operation of such devices [22], as have power frequency fields [23]. Some walkthrough metal detectors employ sinusoidal fields operating at single or multiple frequencies ranging from 300 Hz to 8 kHz, while others use pulsed fields with components up to 70 kHz. Pulsed fields are reported to be a much more significant threat than sinusoidal fields in both [22, 23].

At high frequencies it is not possible to compare the reference levels with the immunity specifications for AIMD. The immunity levels are specified for devices to be tested outside the body, but the in-body fields are not readily related to the external field due to interactions between the field and the body tissues. At low frequencies, however, the body has a negligible effect on the magnetic field, allowing the magnetic field immunity requirements to be directly compared with the reference levels.

Immunity to a magnetic field of 150 A/m RMS over the band 1–100 kHz is required by [21]. This is much higher than the field reference levels of [16] at these frequencies (5 A/m RMS), and higher than the revised ICNIRP general public reference levels of [19] (21–64 A/m RMS). However, traction current transients contain frequencies below 1 kHz [20]. Induced voltage limits are specified in [21] for frequencies down to 16.6 Hz (a railway power frequency). For a representative lead loop area of 230 cm^2, the 2 mV (peak-to-peak) specified from 16.6 Hz to 1 kHz for modulated fields corresponds to magnetic fields ranging from 234.6 A/m RMS at 16.6 Hz to 3.9 A/m RMS at 1 kHz, which approximate the reference levels of [16]. However, the general public limits of [19] exceed those of [16] at frequencies above 25 Hz, reaching 64 A/m RMS at 1 kHz. Thus, the 2 mV requirement corresponds to lower magnetic fields than the ICNIRP 2010 general public limits [19] between 25 Hz and 1 kHz. A further requirement, for immunity to 1 V unmodulated from 16.6 Hz to 167 kHz [21], corresponds to fields that fall below the general public limits of [19] at frequencies above 90 kHz for a 230 cm^2 lead loop, but as low as 54 kHz for a 400 cm^2 lead loop (the maximum size expected).

4 Conclusions

At vehicle level, the EMC test requirements of UNECE Regulation 10 and CISPR 12 have already undergone some adaptation to hybrid/electric vehicles. Nonetheless, further adaptations relating to operating modes will be needed due to the widely differing characteristics of emerging electrical powertrain architectures. Wired charging is now considered in UNECE Regulation 10, although wireless charging has yet to be addressed. However, the adaptation of component level EMC test methods to take account of electrical powertrains is less well developed.

Development of human exposure tests for vehicles has been initiated by standards bodies. However, it should be noted that the interference requirements of EN 45502 for AIMD may correspond to lower field levels than the recently relaxed ICNIRP 2010 reference levels. This, should be considered when setting limits for automotive applications, including wireless charging of traction batteries.

Acknowledgments The research leading to these results has received funding from the European Community's Framework Programme (FP7/2007–2013) under grant agreement no. 314609. The authors are grateful for the support and contributions from other members of the HEMIS project consortium, from CEIT (Spain), IDIADA (Spain), Jema (Spain), MIRA (UK), Politecnico di Milano (Italy), VTT (Finland) and York EMC Services (UK). Further information can be found on the project website (www.hemis-eu.org).

References

1. Accenture (2011) Plug-in electric vehicles. Changing perceptions, hedging bets. Accenture end-consumer survey on the electrification of private transport
2. Sedano B, Ruddle AR, Unanue I, Low L, Astigarraga D, Ibarra I, Cerro I, Galarza A (2013) HEMIS: electrical powertrain health monitoring for increased safety of FEVs. Chem Eng Trans 33:781–786
3. Ruddle AR, Armstrong R (2013) Review of current EMC standards in relation to vehicles with Electric powertrains. In: Proceedings of the 12th international European EMC symposium Bruges, Belgium, pp 298–303, Sept 2013
4. Directive 2004/104/EC of 14 October 2004 adapting to technical progress Council Directive 72/245/EEC relating to the radio interference (electromagnetic compatibility) of vehicles and amending Directive 70/156/EEC on the approximation of the laws of the Member States relating to the type-approval of motor vehicles and their trailers. Off J EU, No L 337, pp 13–58, 13 Nov 2004
5. UNECE (2012) Regulation No. 10—Uniform provisions concerning the approval of vehicles with regard to electromagnetic compatibility, Add 9, Rev 4, 6 March 2012
6. Commission Regulation (EC) No. 661/2009 of 13 July 2009 concerning type-approval requirements for the general safety of motor vehicles, their trailers and systems, components and separate technical units intended therefor", Off J EU, No L 200, pp 1–24, 31 July 2009
7. ISO 11452 Component test methods for electrical disturbances in road vehicles, 11 parts
8. CISPR 25:2002+C1:2004 Limits and methods of measurement of radio disturbance characteristics for the protection of receivers used on board vehicles, 2nd edn. Corrigendum 1, March 2004
9. Janssen V (2011) RFI emission measurements: current status on CISPR standards. www.schurter.com/en/Documents-References/Downloads-Document-Request. Accessed 14 Feb 2011
10. ISO 11451 Road vehicles—Vehicle test methods for electrical disturbances from narrowband radiated electromagnetic energy, 4 parts
11. Schmid G, Überbacher R, Göth P (2009) ELF and LF magnetic field exposure in hybrid- and electric cars. In: Proceedings on bio-electromagnetics conference 2009, Davos, Switzerland, Paper 9, 3 June 2009
12. CISPR 12:2001+A1:2005 (2005) Vehicles, motorboats and spark-ignited engine driven devices. Radio disturbances characteristics. Limits and methods of measurements, 5th edn. Amendment 1, 24 Feb 2005
13. Ruddle AR, Topham DA, Ward DD, Page PJ (1999) Theoretical, analysis of automotive emissions measurements at 3 m and 10 m ranges. In: Proceedings of the 13th Zurich EMC symposium, pp 285–289, Feb 1999
14. SAE Recommended Practice J551-5:2012 (2012) Performance levels and methods of measurement of magnetic and electric field strength from electric vehicles, 150 kHz to 30 MHz, 11 May 2012
15. de Santiago J, Bernhoff H, Ekergård B, Eriksson S, Ferhatovic S, Waters R, Leijon M (2012) Electrical motor drivelines in commercial all-electric vehicles: a review. IEEE Trans Veh Technol 61(2):475–484
16. 1999/519/EC (1999) Council recommendation of 12th July 1999 on the limitation of exposure of the general public to electromagnetic fields (0 Hz–300 GHz). Off J EC, No L 199, pp 59–70, 30 July 1999
17. ICNIRP (1998) Guidelines for limiting exposure to time-varying electric, magnetic, and electromagnetic fields (up to 300 GHz). Health Phys 74(4):494–522, April 1998
18. Directive 2013/35/EU of the European Parliament and of the Council of 26th June 2013 on the minimum health and safety requirements regarding the exposure of workers to the risks arising from physical agents (electromagnetic fields) (20th individual Directive within the meaning of Article 16(1) of Directive 89/391/EEC) and repealing Directive 2004/40/EC. Off J EU, No L 179, pp 1–21, 29 June 2013

19. ICNIRP (2010) Guidelines for limiting exposure to time-varying electric and magnetic fields (1 Hz–100 kHz). Health Phys 99(6):818–836
20. Ruddle AR, Low L, Vassilev A (2013) Evaluating low frequency magnetic field exposure from traction current transients in electric vehicles. In: Proceedings of the 12th international European EMC symposium, Bruges, Belgium, pp 78–83, Sept 2013
21. CENELEC, EN 45502 Active implantable medical devices, 2 parts
22. Kainz W, Casamento JP, Ruggera PS, Chan DD, Witters DM (2005) Implantable cardiac pacemaker electromagnetic compatibility testing in a novel security system simulator. IEEE Trans Biomedical Eng 52(3):520–530
23. della Chiara G, Primiani VM, Moglie F (2007) Experimental and numeric investigation about electromagnetic interference between implantable cardiac pacemaker and magnetic fields at power line frequency. Ann Ist Super Sanità 43(3):248–253

Advanced Electronic Architecture Design for Next Electric Vehicle Generation

Ovidiu Vermesan, Mariano Sans, Peter Hank, Glenn Farrall, Jamie Packer, Nicola Cesario, Harald Gall, Lars-Cyril Blystad, Michele Sciolla and Ahmed Harrar

Abstract Future generations of electric vehicles (EVs) require a scalable, layered architecture addressing different system aspects such as scalable modules, uniform communication, and hardware (HW) and software (SW) architectures. This will

O. Vermesan (✉) · L.-C. Blystad
SINTEF, Forskningsveien 1, 0314 Oslo, Norway
e-mail: Ovidiu.Vermesan@sintef.no

L.-C. Blystad
e-mail: Lars-Cyril.Blystad@sintef.no

M. Sans
Continental, 1 Avenue Paul Ourliac, 31100 Toulouse, France
e-mail: Mariano.Sans@continental-corporation.com

P. Hank
NXP Semiconductors, Stresemannallee 101, 22529 Hamburg, Germany
e-mail: Peter.Hank@nxp.com

G. Farrall · J. Packer
Infineon Technologies, Great Western Court, Hunts Ground Rd, Stoke Gifford, Bristol BS34 8HP, UK
e-mail: Glenn.Farrall@infineon.com

J. Packer
e-mail: Packer.external@infineon.com

N. Cesario
ST Microelectronics, Strada Primosole, 50 95121 Catania, Italy
e-mail: Nicola.Cesario@st.com

H. Gall
Ams, Tobelbader Straße 30, 8141 Unterpremstätten, Austria
e-mail: Harald.Gall@ams.com

M. Sciolla
Centro Ricerche Fiat, Strada Torino 50, 10043 Orbassano, TO, Italy
e-mail: Michele.Sciolla@crf.it

A. Harrar
PSA Peugeot Citroën, 75 Avenue de la Grande Armée, 75116 Paris, France
e-mail: Ahmed.Harrar@mpsa.com

B. Müller and G. Meyer (eds.), *Electric Vehicle Systems Architecture and Standardization Needs*, Lecture Notes in Mobility,
DOI 10.1007/978-3-319-13656-1_8

reduce the number of electronic control units as well as the variety of communication, sensor data fusion and charging infrastructure interfaces. The architecture is based on distributed processing with novel propulsion systems and electronic control units implemented as embedded systems containing HW and SW algorithms. Sensing, actuation, signal processing and computing devices are embedded in the electronic equipment, motors, batteries and the mechanical components. This paper presents the current advances in novel EV architectures based on embedded computing devices, communication systems and management algorithms.

Keywords Electric vehicle · Electronic/electric architecture · Networking architecture · Partial CAN · FlexRay · Ethernet · Multi core microcontroller · ADAS · X-by-Wire control · Parking assistance

1 Vehicle Architecture and Systems—Domain Partitioning

The overall EV architecture proposed in the ARTEMIS POLLUX project is designed as a multi-layered hierarchical structure with domains having different boundaries as presented in Fig. 1. The EVs and hybrid electric vehicles (HEVs) present unique challenges when it comes to the electrical/electronic (E/E) architecture which is becoming more and more complex and includes new functions. In this context, new concepts are needed to enable an easy introduction of new technologies along with an interconnected but still manageable architecture. The EV architecture design needs to focus on optimizing costs, providing peak performance and offering robust EV components and systems including:

- Charging Systems
- Connection Systems
- Electrical Centers
- Energy Storage Systems
- Power Electronics
- Powertrain Systems
- Service Solutions
- Thermal Systems
- Wiring Assemblies

An efficient and robust structured system architecture design is necessary more than ever due to the rising complexity of functions in next generation EVs (Fig. 2). The E/E system architecture design focus in the ARTEMIS POLLUX project was on the following elements:

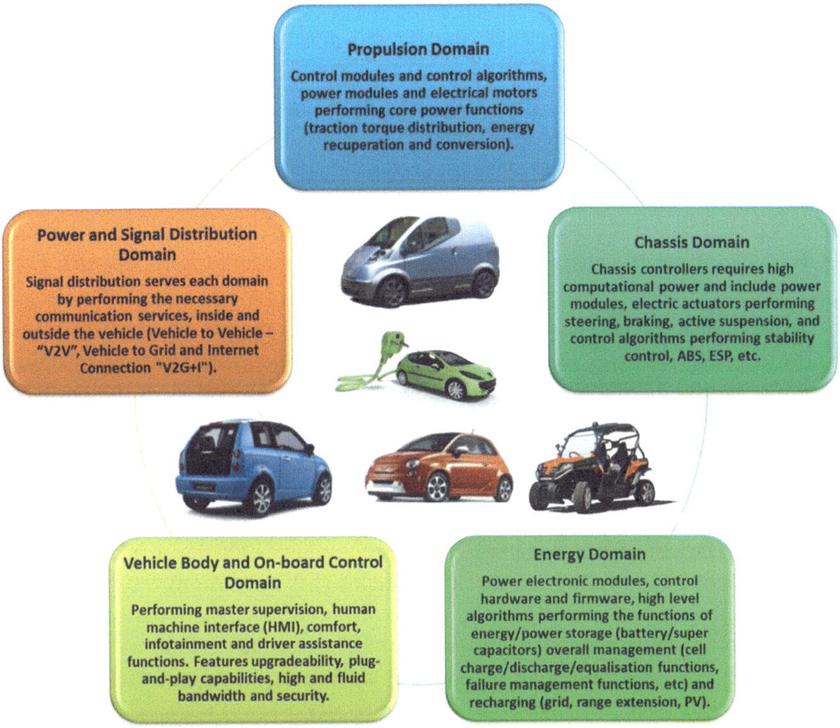

Fig. 1 EV architecture domain partition

- Partitioning of functions
  - Communication network (controller area network (CAN), FlexRay, Ethernet)
- Computing and processing using multicore microcontrollers
- Energy management
- Propulsion and core power functions
- Power and signal distribution
- On-board control, X-by-Wire and human machine interfaces (HMIs)
- Effective variant coverage

1.1 Architecture Concept Description

Electric propulsion technologies are offering to significantly improve vehicle performance and energy utilization efficiency [1]. Sub-systems such as the electric propulsion system, electric brake- and drive-by-wire systems have certain measures for mitigating potential hazards which could occur due to both random failures and systematic failures. These measures include fault-tolerance including fail-safe, fail-operational capabilities. ISO 26262 [2] is a new standard for the functional safety of E/E systems in road vehicles. The guidelines for the safety management process for developing such measures are used in the ARTEMIS POLLUX project to develop the architecture and embedded systems needed for the next generation EVs.

1.1.1 Layered Architecture

EV architectures need more HW/SW integration and require moving from pure hardware to mechatronics systems and intelligent modules. Future EV architectures will move in the direction of virtualisation with a transformation of the hardware architectures into scalable functional architectures. In this context there is the need for architecture layering and the introduction of this concept both on the level of the EV (electronics, communication, energy management, software layers) and as well on the individual layers in the architecture (i.e. communication layer with Ethernet backbone). The concept is presented for the communication layer and opens the way to implement the future Internet of Vehicles (IoV) paradigm.

The layering concept enables the move from off the shelf hardware to off the shelf functionality embedded in the layered architecture. This will allow the development and distribution of automotive software applications that create the

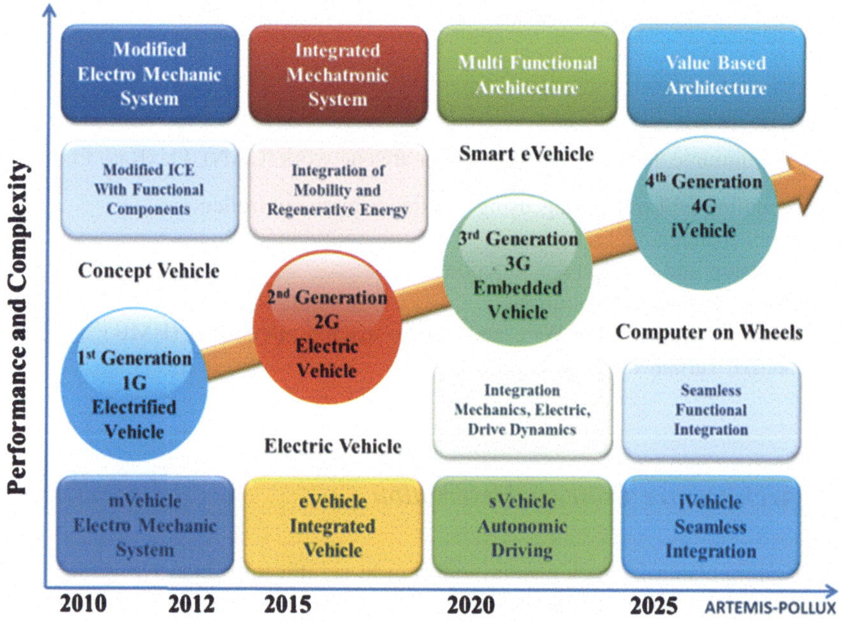

Fig. 2 EVs generations

possibility of providing new information and communications technology (ICT) architecture that enables to adapt the HW based functionality into the EV over the vehicle life cycle. The layered architecture concept offers a reliable supply of electrical energy for the electrically controlled and powered systems such as braking, steering, stabilisation, etc. and an intelligent integration of the storage medium into the overall concept of the vehicle energy management architecture.

2 Modelling and Simulation Platform for EV Architecture and System Design

The design of a full EV requires the development and optimization of a complete electric powertrain, including battery, power electronics, electric machine, sensors and control system. The ARTEMIS POLLUX project approached the designing of the electrical platform, using V development cycle and proposed a modelling and simulation platform in order to drive the main choices and then to optimize the system. Physical testing is used extensively to characterize mechanical systems, but in many cases mathematical models are available that adequately describe the behaviour of parts of the system. Thus, test systems can consist of either only physical parts, both physical and virtual parts, or only virtual parts. In the case of both physical and virtual parts one has a real time computer simulation, and has the potential to enhance convenience and reduce costs. The term model-in-the-loop (MIL) is used by the automotive industry to describe this concept. Real time simulations of vehicle dynamics were performed in the project. Vehicle and powertrain models were integrated into a MIL environment. They were verified by use cases within drive cycles derived from real-life measurements. The simulation environment was built around the AVL CRUISE, MATLAB/SIMULINK and CRUISE Model Compiler. Computing and Processing in Electric Vehicle. The EV layered architecture proposed by the ARTEMIS POLLUX project requires advanced computing power in the vehicle provided by using high-performance multicore processors and novel virtualization technology software concepts and operating systems. The design of the novel EV architecture focuses on reducing the size, weight, and complexity of the hardware used in EVs and at the same time offering more and more electronic features through the use of multicore processors (processing chip with several 2–4–8 processing units). The approach considered by the ARTEMIS POLLUX project was to use integrated systems on chip based integrated multicore processors that are in line with the proposed domain architectures considering the following elements:

- Separation of functions and safety critical support in such a way that different functions are able to run simultaneously without affecting each other. The safety critical functions run in parallel with non-safety-critical functions without their safety characteristics being compromised.

- Support and integration of multiple operating systems because different functions are covered by different operating systems (e.g., AUTOSAR (AUTomotive Open System ARchitecture) for safety-critical functions, Linux for automotive vehicle controller/HMI, QNX for HMI, and Android for Vehicle-to-Infrastructure (V2I) and user apps) and it is necessary that the multicore system is flexible and runs multiple operating systems at the same time.
- Efficient shared use of multicore system resources since different functions use some of the same dedicated system resources (shared use of communication channels, share use of processing for multi propulsion).

2.1 Controller Architecture

The main microprocessor architecture work in the POLLUX project was to improve support for multicore controllers running multiple applications. For safety-relevant applications in automotive, this requires mechanisms to protect the different cores and software from interfering with one another.

It is, in principle, possible to use any standard microprocessor in a safety-critical system by employing fault-detection techniques such as redundancy and diversity at the system level. However, the costs and complexity of doing this will be larger than using a processor with safety features designed in. Using external hardware to monitor and detect faults will also have a longer reaction time than hardware within the processor core. To provide some context, the project reviewed the support for safety-critical systems in some commonly used embedded processors and FPGA products. This was not an exhaustive survey but an overview of representative technologies including ARM, LEON (SPARC), MicroBlaze, NIOS II, PowerPC and FPGA solutions. Some of these do have limited support for fault tolerance (mainly soft error detection and correction in memory and registers). However, none appeared to meet all the requirements identified by the project, particularly in terms of support for multi-processor systems (e.g. checking of all data communication within and between cores) and software encapsulation (protection of memory and registers from erroneous accesses).

No single architecture can meet all users' needs, for both technical and commercial reasons. Therefore, two alternative microcontrollers that meet the requirements of the project, but based on different processor cores, are presented. Both microcontrollers provide a complete System on Chip (SoC) solution containing program and data memories, buses, bus arbitration, interrupt system, multiple direct memory access (DMA) controllers and a powerful set of on-chip peripherals such as serial controllers, timer units, analogue-to-digital converters and network communication interfaces. They are designed to meet the needs of safety-critical applications where the competing issues of price, performance, real-time responsiveness, computational power, data bandwidth, and power consumption are key design elements.

These two architectures take different approaches to meeting the requirements for scalable safety, performance and power, monitoring and managing reliability failures due to aging, and protection mechanisms. They make different design trade-offs and design decisions to meet the requirements.

2.1.1 Automotive Microcontroller TriCore Architecture

The microcontroller architecture is based around a new version of Infineon's TriCore architecture [3] with extensions for improved protection, performance and power. This enables a single architecture to support powertrain, body and safety domains. The architecture provides scalable safety, scalable performance and scalable memory systems; all based on a common safety case. A typical implementation of such a controller, as shown in Fig. 3, uses three TriCore processors, program and data memories, buses and arbitration, interrupt system, multiple DMA controllers and a set of on-chip peripherals. This is designed to meet the needs of the most demanding automotive controller applications, meeting the competing issues of safety, price/performance, real-time responsiveness, computational power,

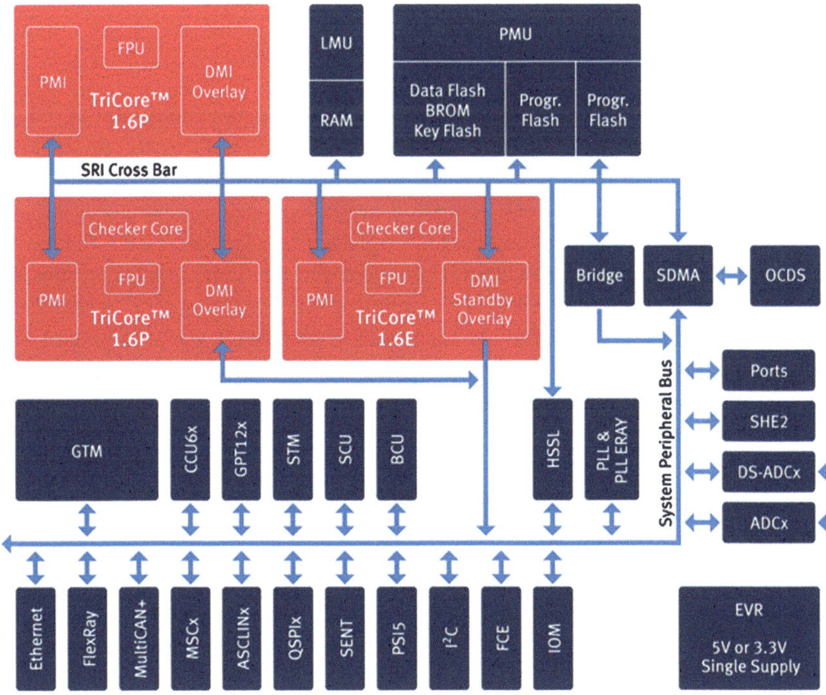

Fig. 3 Controller architecture

data bandwidth, and power consumption. The main goals of the POLLUX project in the development of the new TriCore TC1.6E processor architecture were:

- Multicore support: Smaller, lower power core. Instruction set enhancements and improved interrupt system. Replace data cache with data buffer to remove cache coherency issues.
- Support for Automotive Safety Integrity Level (ASIL) D: Lock-step checker cores provide redundancy and fault coverage for both permanent and transient faults, with no loss of performance.
- Data integrity: Improved support for reliable communications and data integrity checking (CRC calculation, etc.).
- Software encapsulation: System level protection mechanism for memory and registers. Extended protection with password-protected access to safety-critical system registers.
- Reduced power consumption: Scalar core optimized for energy efficiency with dynamic power control: idle, sleep and standby modes, clock control for CPUs and peripherals. Removal of data cache and simplified memory system.
- Automotive Open System Architecture (AUTOSAR) support: Basic Software ported to take advantage of new safety features.

2.1.2 CPU Safety Features

Two of the key challenges were to provide lock-step redundant cores and to reduce power. The use of a scalar core helps in both areas. This reduces the power consumption (and size) of the core itself by reducing the complexity. It also reduces the number of signals that need to be compared in the lock step comparator logic and the number of flip-flops needed to delay the signals. Each TriCore 1.6E CPU has an identical checker or shadow CPU that is provided with the same inputs and runs the same instructions as the main CPU. The outputs of the two cores are compared and any errors reported to the Safety Management Unit (SMU) via a redundant signalling mechanism. This provides single point fault metric >99 % and latent fault metric >90 % for both permanent and transient faults as required to meet ASIL D.

2.1.3 Support for Software Encapsulation

Writing and developing safe software is a demanding and costly undertaking. For this reason it is very desirable to encapsulate any safety-relevant software, so that it has a guaranteed time budget, schedule and memory resource. Encapsulating all the software modules enables software modules to be developed according to their own safety integrity levels. Only common system components have to fulfil the highest ASIL requirement. This leads to reduced development costs and ensures a higher system safety and stability.

2.1.4 Automotive Dual-Core Microcontroller HW Platform

A hardware (HW) motor controller hardware platform was developed by STMicroelectronics S.r.l. The HW platform has to be adaptable to manage different sizes of power. It has to be scalable from master to slave configuration. The HW platform has to guarantee the ASIL C/D required for the application. In order to meet these requirements a safety dual-core microcontroller was developed, based on the Power Architecture™ cores. It implements the following features:

- A "unique scalable" multi-core safety core architecture
- Innovative safety concept matches ISO 26262 ASIL D
- On-chip redundancy provided for critical components of the microcontroller
- 5-stage pipeline dual issue core provides a high efficiency
- Lock step Redundancy Checking Units are implemented at each output of the Sphere of Replication (SoR)
- Programmable fault collection and control unit monitors the integrity status of the device and provides flexible safe state control
- Two modes of operation: LSM (Locked Step Mode) and DPM (Decoupled Parallel Mode)
- Single or double read for all analog input
- Single or double read for position sensors
- External watchdog function
- Power supply monitor function
- Error out monitor function (FCCU)
- Safe relay management

3 Networking in Electric Vehicle

Communication and bandwidth requirements increase as more new and complex applications appear in the vehicle, for example, enhanced safety and entertainment solutions. End users expect the same level of entertainment functions and data in the vehicle as known from home.

Communication standard ISO 15118 is one of the central standards for communication between EVs and the Electric Vehicle Supply Equipment (EVSE) [4]. It describes the communication between the EV communication controller and the supply equipment communication controller. Thus implementing the ISO 15118 enables a standardized communication method for multiple of applications and services. It can be related to e.g. identification, authentication and authorization during charging of EVs, and the charging process itself.

3.1 Power and Signal Distribution—In-vehicle Network Architecture

Existing vehicle control networks, based on the Local Interconnect Network (LIN), CAN, and FlexRay standards, are not designed to cover these increasing demands in terms of bandwidth and scalability that we see with various kinds of Advanced Driver Assistance Systems (ADAS).

For in-vehicle networking, Ethernet provides all the pre-requisites for a holistic approach. It is suitable as backbone bus to connect the various application domains as well as within domains, when higher bandwidth is required. Today, switched Ethernet networks base on point-to-point communication and available bandwidth can more efficiently used compared to broadcast systems like CAN or FlexRay. While previous generations focus on a certain application domain, generation 3 of automotive Ethernet will introduce the backbone of the in-vehicle network. A typical backbone is illustrated in Fig. 4. Such network architecture introduces a new concept for how communication is organised between the ECUs and the network management. Needless to say that software content increases while communication is structured in a hierarchical way with domain controllers connected via Ethernet backbone and switches.

This structure represents a scalable solution as each port of a switch can be implemented as 10 Mbps, 100 Mbps, or 1 Gbps without any change in higher protocol layers. The paradigm shift is also visible in how a message is transported via domain boundaries to its destination. While in a vehicle network today complex and network-dependent gateways realise the data transport, existing and matured IP-based routing concepts within switches and routers are proposed for the

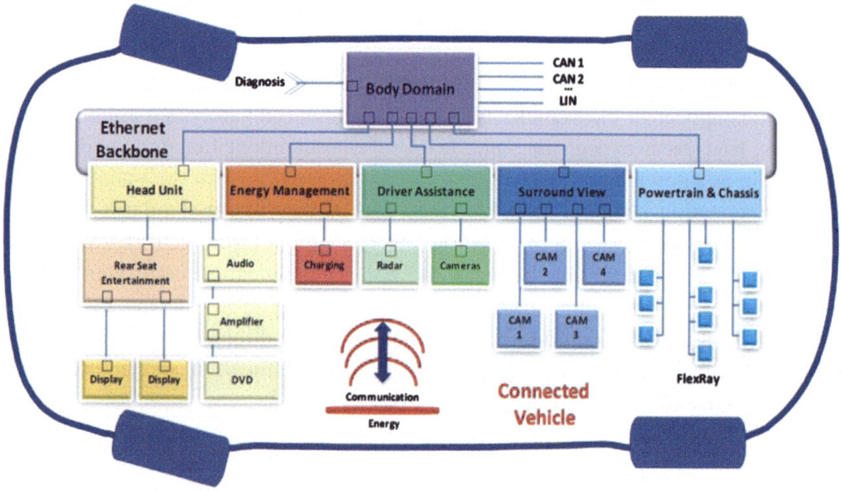

Fig. 4 Ethernet backbone in domain architecture

backbone network. The advantage is that the IP-based routing does not depend on the underlying network implementation, thus allowing a unified addressing concept for the whole in-vehicle network. Moreover, the IP-based routing enables the straightforward connection of the vehicle to the Internet [5], a trend which is mainly driven by end users who expect the same access to services as in office and home environments. A further characteristic of the new style of architectures is the usage of one and only one backbone network technology, namely Ethernet, which has to accommodate different data communication classes like diagnostics, video/audio streaming, and highly dependable control data.

While Audio/Video Bridging (AVB) [6] Ethernet and Time Triggered Ethernet [7] can already provide different levels of Quality of Service (QoS) in combination with real-time performance, further work is needed to validate and ensure the coexistence of different data classes on the same communication structure. In addition, encryption of data is required to secure communication with the infrastructure. Figure 5 gives an example for a data flow through different layers of the Ethernet stack. Data packets pass 100BASE-TX and Unshielded Twisted Single Pair (UTSP) Ethernet networks, and, finally, the wireless link to the infrastructure (grid).

Electric and hybrid vehicles are entering the automotive market. Their presence on the roads is expected to rapidly grow over the next decade. In-vehicle networking and infrastructure (grid) communication are important aspects of EV functions. EV applications impose considerably higher requirements on electronic systems and their components in terms of Electromagnetic Compatibility (EMC) [ISO 11452] and environmental conditions.

Initial investigations show that BroadR-Reach technology for the Physical Layer is suitable for its usage in the automotive environment. However, to achieve the necessary robustness for a next-generation vehicle networking standard, new optimized components have to be developed. The system diagram in Fig. 6 reveals

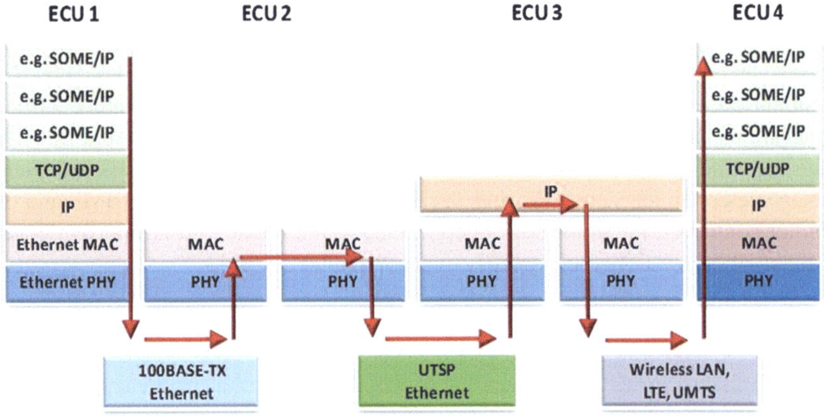

Fig. 5 Data flow through layered architecture

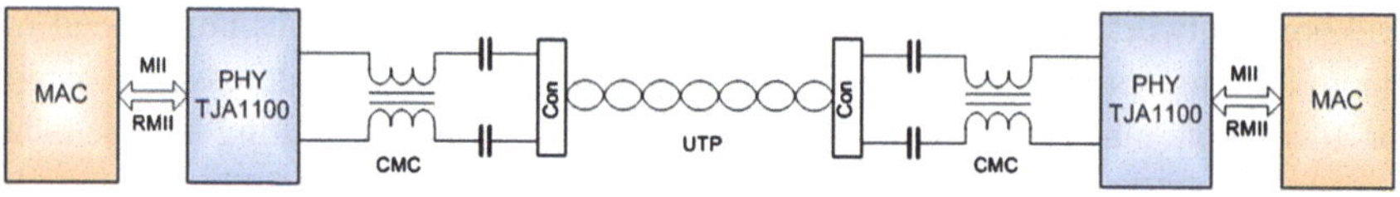

Fig. 6 BroadR-Reach system diagram

the main components of the automotive BroadR-Reach link. Compared to 100BASE-TX (Fast Ethernet), the bill of material is significantly reduced.

A physical layer component, e.g. TJA1100, NXP [8], as the interface between the analogue transmission medium and the digital Medium Access Controller (MAC), largely determines the robustness and emission performance of the link. While a consumer physical layer is optimized to support a cable length of more than 100 m, automotive solutions have typically to deal with a link length of less than 10 m.

Using a backbone in a domain based architecture with high-speed communication results in less Electrical Control Units (ECUs) and less cables. For all vehicles, especially the EV, the benefit of smaller cable harness is cost, weight, and energy consumption. BroadR-Reach [9] technology allowing the usage of unshielded twisted-pair cable and makes Ethernet cost-competitive for automotive applications.

3.2 Enhanced FlexRay^TM^ Standard Transceiver

One of the vehicle demonstrators in the POLLUX project was a CRF prototype A-class PHEV. The communication among the vehicle management unit (VMU), the motor controller and the battery management system (BMS) is based on the FlexRay bus. Each unit communicating through the FlexRay bus is equipped with ams AG AS8222 Transceiver. It features improved immunity (BCI, DPI), improved ESD and increased temperature operation range for application at engine/motor condition (from −40 up to +160 °C). It incorporates FlexRay V2.1 Rev B Bus-interface, and bus and local wake-up detector for power management. The transceiver has hardwire microcontroller interface for power modes, FlexRay communication controller interface, bus guardian interface for safety critical applications and power supply interface for managing external power supplies. The state machine features normal, receive-only, standby and sleep mode, and under voltage detection for V_{BAT}, V_{IO} and V_{CC}. For the demonstrator, in the BMS and the eMotor controller, the transceiver is connected to the FlexRay controller integrated in the dual core STMicroelectronics microcontroller.

4 Energy Management in Electric Vehicle

Energy management is key to ensure that the EV and the charging infrastructure are safe, convenient, economical, and energy efficient for drivers and facility managers, while ensuring that utility operators are able to manage the impacts of EV charging on the grid. Efficient energy management system will allow linking massive numbers of EVs to the grid, opening the way to implementing the future Internet of Vehicles (IoV) paradigm and creating electro mobility ecosystems by integrating different stakeholders (consumers, facility/vehicle fleet managers, and electric utilities).

4.1 Battery/Supercapacitors Monitoring System

Combining batteries with other storage devices, such as super capacitors allows the inclusion of power electronic converters at the storage energy level and make possible to decouple the power (acceleration, braking mode) and energy (cruise speed) functions of storage, providing lower power levels in batteries, and improve the energy management efficiency in the storage system [10] [11]. A battery/supercapacitors monitoring system (BSMS) was developed in the project. Its monitoring functions were integrated in the overall system design. The dependency between the monitoring functions and the rest of management functions and safety checks are considered one single functional unit that, as a result, manages the battery pack in a safe and optimized way. The software components included in the BSMS system were implemented following the AUTOSAR description, and are thus considered SWC as described by the standard. The main software module responsible of the BMS functionality control is the SWC working mode manager module. It belongs to the application layer and it is based on a communications interface layer which permits the communication with the rest of modules and the others ECUs connected to the system (i.e. intelligent sensors and actuators). The main role of the working mode component is to identify the operation mode of the BSMS and issue open/close commands for the different relays in BMS accordingly. The relays present in the BSMS system are:

- Positive and negative battery relays
- Charger relay
- Pre-charge Charger relay
- Inverter relay
- Pre-charge Inverter relay
- Auxiliary compressor relay
- Auxiliary DC–DC-1 and DC–DC-2 relay

The operation modes considered are: OFF (low consumption cycle), INIT, RUN, CHARGE, RECOVERY, EMERGENCY (Failure mode with safe-state condition). Each operation mode issues a different combination of relay commands that configure the power-flow path inside the power distribution box. Transitions between states are controlled by two types of inputs to the model: The internal signals are generated by other SWC in the system affecting the working mode and therefore the action to the power relays. The external signals are generated from external ECUs. The other main software components of the BSMS include:

- SWC Relay Control manager and Diagnostic module
- SWC Charge Manager
- SWC Discharge Manager
- SWC Energy Balancing Control
- SWC Thermal Manager

The BSMS is equipped with smart cell sensors and actuators, battery front-end sensing actuators. The overall goal is to have a safe battery management system. The BSMS is being continuously monitored to detect any errors or critical conditions. In case of any safety violations, appropriate action is being taken. The level of safety functions are basically split into six different modules:

- CAN Diagnostic Monitoring
- Pack Current Monitoring
- Cell Temperature Monitoring
- Cell Voltage Monitoring
- Isolation Fault Monitoring
- Fault Reaction Monitoring

4.2 DC–DC Converter Module

A DC–DC converter targeted for use in an EV was developed in the project. It is intended to supply the low voltage battery and the related auxiliary loads (Table 1). The full bridge phase-shifted zero-voltage-switching architecture used in the DC–DC converter is based on Infineon power MOSFETs OptiMOS 250 V. The H-bridge is packed in an Easy1B package, to optimize matching, minimize loss, and ease the PCB mounting. Nominal input voltage for the DC–DC converter is the high voltage battery voltage (96 V), which is then down-converted to 13 V. The DC–DC converter is designed with partial CAN bus communication module. Using the CAN bus, the motor controller can power-up and power-down the DC–DC converter, and monitor its current consumption and the ambient temperature.

Figure 7 shows the block diagram of the DC–DC converter with the main blocks included. At the core, the DC–DC module includes Infineon's Full-Bridge Easy1B module with OptiMOS^{TM} power MOSFET's.

The CAN module is controlled by an 8-bit ATMEL microcontroller, CAN 2.0A/B and ISO 16845 Certified. The CAN bus is controlled by NXP TJA1145 high-speed CAN transceiver for partial networking. It is Fully compatible with the ISO 11898-2, ISO 11898-5 and ISO 11898-6 standards. The partial networking function makes it possible with a remote wake-up via standard wake-up pattern or via ISO

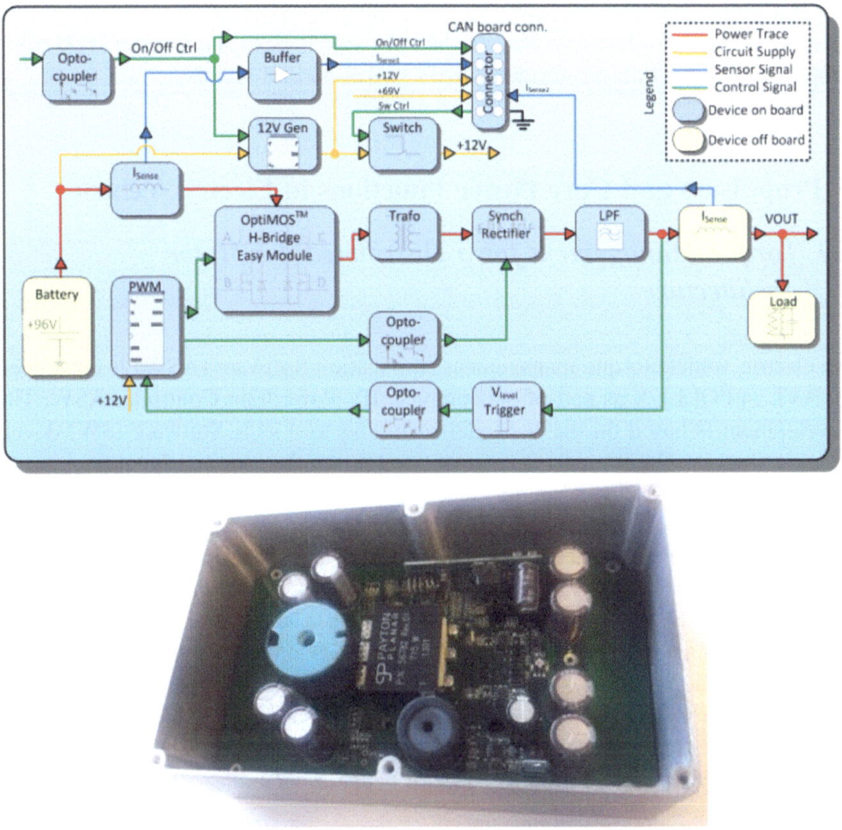

Fig. 7 DC–DC Converter architecture and implemented prototype

Table 1 DC–DC converter specification

Parameter	Value
Topology	Full bridge, phase-shifted w/synchronous rectification
Networking/communication	Partial CAN (ISO 11898-6)
Efficiency	>90 %
Switching frequency	100 kHz
Input voltage (nom.)	96 V
Output power (nom.)	400 W
Output voltage range	12.2–13.8 V
Output current (nom.)	20 A
Galvanic isolation	Yes

11898-6 compliant selective wake-up frame detection. The DC–DC converter with the CAN module is shown in Fig. 7. The enclosure is designed to meet IP66 when properly installed with pre-formed silicon gasket and lid screws.

5 Propulsion and Core Power Functions in Electric Vehicle

5.1 Torque Management for a Distributed 4 e-Motor Architecture

An electric vehicle torque management Application Software (ASW) is developed by AVL in POLLUX as part of the complete EV Powertrain Controller ASW. The development is based on the V-Model according to AVL's Software (SW) development process. The torque management comprises the required functionalities to calculate electric machine torque request as a result of the driver interaction. It considers torque relevant driving functions, and a torque vectoring controller prototype software for a distributed 4 e-motor architecture. Promising results regarding improvement of dynamic vehicle behaviour was shown. With the dynamic torque split among the 4 e-motors the average speed in a slalom manoeuver could be increased by almost 7 %. The e-Torque Vectoring Controller enables furthermore higher lateral acceleration and improves handling and agility of the vehicle. The steering support achieved by torque vectoring functionality reduces the steering effort and thus contributes to improved vehicle controllability.

Dynamic torque distribution has several benefits, including

- Faster acceleration, no sliding wheels
- Higher efficiency due to higher recuperation torque/power utilization
- Increased safety at braking/recuperation manoeuvres
- Improved sportiness by enhanced vehicle lateral dynamics
- Improved turn in behaviour at steering into a bend
- Increased EV stability by actively influencing the EV yaw behaviour
- Increased EV safety in curves by optimized use of the traction potential
- Extended range or potential to reduce battery size
- Improved drive off at different road surface conditions

A simulation environment has been used for development, optimization and calibration of the e-Torque Vectoring Controller (eTVC) as well as for demonstration of the vehicle dynamics improvements and goal achievements. The closed loop simulation tool chain consists of the e-Torque Vectoring Controller and the vehicle simulation model as plant model. The vehicle plant model is built with the combination of vehicle dynamics model in CarMaker and the powertrain model in AVL CRUISE. The eTVC prototype software has been developed in Matlab®/Simulink® and is connected to the plant model with a defined interface.

6 On-Board Control, X-by-Wire and Human Machine Interfaces

6.1 X-by-Wire Parking Assistance System

An innovative semi-automatic X-by-Wire Park Assistance system has been designed in the ARTEMIS POLLUX project by Continental Automotive France and integrated in cooperation with ZEM AS in an EV demo vehicle with 34 kW traction power and a 24 kWh battery that provides 160 km range. This system is based on a handy joystick-driven actuation, able to control simultaneously steering, gear and traction engine during parking manoeuvres in a semi-automatic way. Successive turnings front and rear are automatically programmed, so that the driver doesn't need to repeatedly actuate the pedal, gear selector and steering wheel during the manoeuvres. The Cartesian XY position of the joystick defines a 'vector' control, used to program and adjust steering direction, speed and amplitudes of the vehicle movements with a high accuracy during the parking manoeuvres, facilitated by the fast, secure and accurate controllability of electric type motors.

Parking assistance systems are currently being proposed by vehicle-makers, as one of the most original ADAS equipment to be frequently integrated in future vehicles.

Passive systems only propose a set of sensors to detect available space and obstacles, typically with move back radars, to alert the driver during parking manoeuvres.

Active or semi-active systems can control the steering wheel following a pre-defined trajectory calculated from the free area detected by spatial sensors, either in a full automatic mode, or in a semi-manual mode requiring the driver to shift the gear and push on the accelerator pedal [12–15].

The solution developed in the ARTEMIS POLLUX project is a semi-automatic manoeuvre control system, activated via an X-by-Wire Joystick action from the driver, able to facilitate difficult manoeuvres, without any external sensor. Main target is to help the driver for low speed manoeuvres needing both traction and steering controls, to avoid acting manually on the accelerator pedal, brake pedal, gear and steering wheel, during the successive repeated and annoying 'push-&-pull' or 'front-&-rear' steps, as on typical parallel parking manoeuvres. The system was installed on the EV using fast prototyping tools (Matlab and dSPACE).

6.1.1 General Description

This X-by-Wire assistance system is based on control by joystick allowing the driver to pilot the vehicle without any action on pedals, gear, or steering wheel.

For cost and complexity reduction reasons, no external sensors for available distance and space measurement and obstacles detection are proposed; the driver is supposed to check the displacements on mirrors to keep the control of the vehicle movements, and is able to brake at any moment, to maintain safety. For this reason this system is to be considered as a "semi-automatic" parking assistance system.

The control can be done by a stick (as seen in Fig. 9) or a trackball, or any other equivalent 2- or 3-axes handy-actuator (all named in this document with the generic word "Joystick"), to be used in an intuitive way by the driver to indicate the desired vector direction and amplitude, which are converted into a steering angle set-point and into a traction torque or speed set-point.

This semi-automatic moving control builds a trajectory of the steering wheels so as to follow this displacement vector, as described further. In contrast, as soon as the Joystick is reset to its central (neutral) position, the vehicle traction is stopped and braked even on a sloped road. The direction target corresponds to the final displacement of the vehicle to get it parked on its final position as desired (Fig. 8):

Either the trajectory can be a simple one, or a combination of short movements. In that case, this HMI solution can offer a significant advantage, as explained further. The vehicle control architecture, showing the connection of the Joystick to

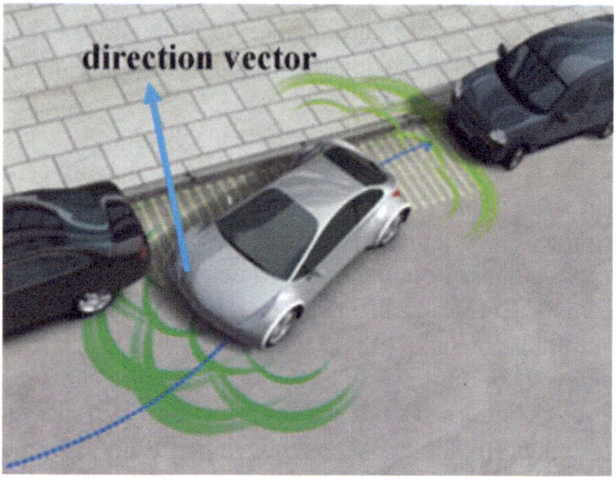

Fig. 8 Trajectory to get final parking position

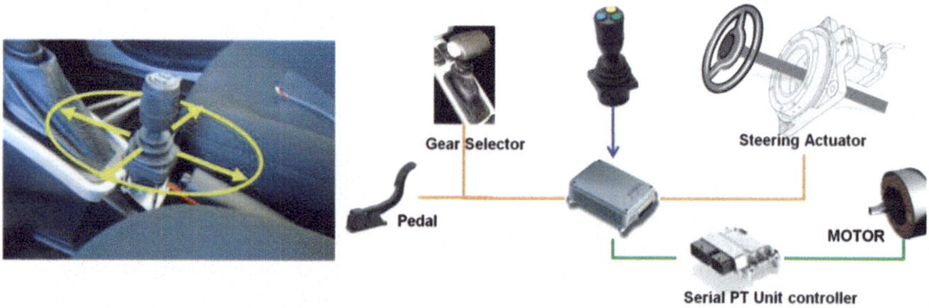

Fig. 9 The vehicle control architecture

a supervisor controller via the CAN data network, to control both powertrain engine and power steering system is presented in Fig. 9. This X-by-Wire solution can be adapted on any vehicle equipped with an automatic gearbox and a steering assistance system, but having any type of powertrain technology: conventional ICE, pure electrical or hybrid.

The new solution developed in the ARTEMIS POLLUX project takes advantage of the capability of EV and doesn't need any mechanical multi-ratio gearbox in their powertrain configuration, thanks to the reversibility and symmetry of electrical engines, able to run in any rotation and torque directions with fast inversions.

No mechanical shifting is required and even changing from rear to front and vice versa can be managed electronically by software.

Full automated control of the torque and speed of traction remains open for soft, automated and fast control, ideal basis for the parking assistance system implemented.

6.2 Human Machine Interface (HMI)

A HMI demonstrator has been developed in the project by City Motion and SINTEF, and includes an instrumentation display/vehicle controller. The demonstrator consists of two main units; an LCD display and the controller unit. Its logical architecture is shown in Fig. 10. The controller unit is a single board computer based on and ARM CPU.

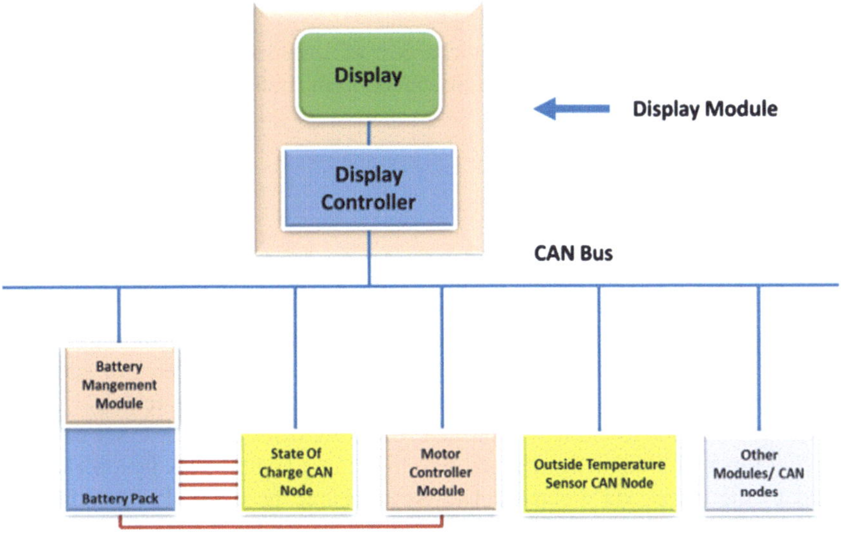

Fig. 10 Display/vehicle controller—logical architecture

Fig. 11 HMI display design

The display controller displays various pieces of information on the graphical display units based on the information sent by peripheral CAN modules over the CAN bus. The vehicle/display controller unit is capable of running several different operating systems. In our work, we chose two different operating systems, namely Linux and QNX, and developed two main versions of the demonstrator based on these operating systems.

Display visualisation is highly dynamic based on the information flowing through the CAN bus. It has a state dependent look-and-feel, for example at a lower vehicle speed the theme colour is green whereas at speeds higher than a threshold it changes to red. The vehicle controller displays various pieces of information on the graphical display units. These pieces of information which are originated in other units such as the Motor Controller Unit (MCU) and the Battery Management System (BMS), are transmitted to the Display Controller over the CAN bus. A partial CAN module was implemented as well for the battery management system. The graphic design is shown in Fig. 11.

Such information includes State of charge (SoC), vehicle speed, the energy used, outside temperature, motor temperature, battery temperature, motor related alarms, and body related warning message.

6.3 Haptic Pedal HMI

A demonstrator based on Democar Quicc Diva vehicle from DURACAR Holding Company has been developed in POLLUX. The principle of this demonstrator is to present Human Machine Interface by using a Haptic Pedal to control gathering speed. An active force feedback pedal has been integrated in the vehicle.

The pedal is connected to a dedicated Micro AutoBox through CAN network. Specific function is implemented and developed in a control desk laptop and provided to the driver via a screen. The Active Force Feedback Pedal (AFFP) is developed by Continental and is an active accelerator pedal. It can change its force versus angle characteristic line and thus the counterforce sensed by the driver at

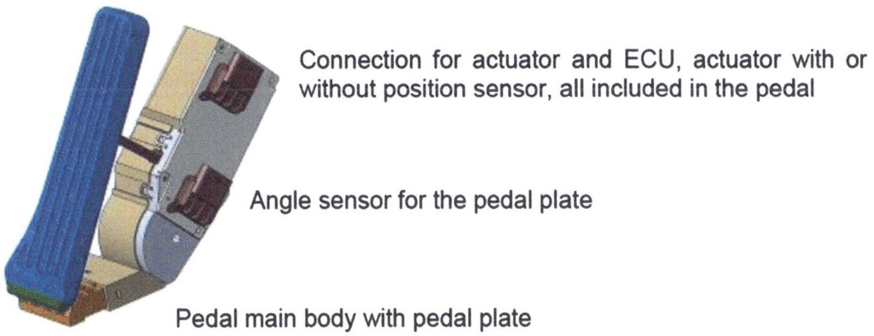

Fig. 12 Components of an accelerator force feedback pedal

actuating the pedal. If this change takes place target-oriented, the driver can be influenced in his driving behaviour.

When operated passively, the AFFP functions like a conventional accelerator pedal. Its basic structure consists of four components, both at standing and at suspended accelerator pedal (Fig. 12).

It is the purpose to use the AFFP pedal to stimulate the driver to drive safely and efficiently. As a whole the AFFP can be regarded as a driver assistance system.

The driver is guided by his accelerator pedal and warned in dangerous situations. However he always has the full control over the accelerator pedal, since, if necessary he can bring it in the ultimately desired position with consciously applied, additional effort. A dashboard display is used to provide the driver with legally required information and driving relevant information, see Fig. 13.

The information is shown in such a way that the driver is not distracted from his driving task, and presents the symbols and the data on the dashboard on a visually comfortable way. So the driver is able to react with a minimum response time which is needed in case of emergency.

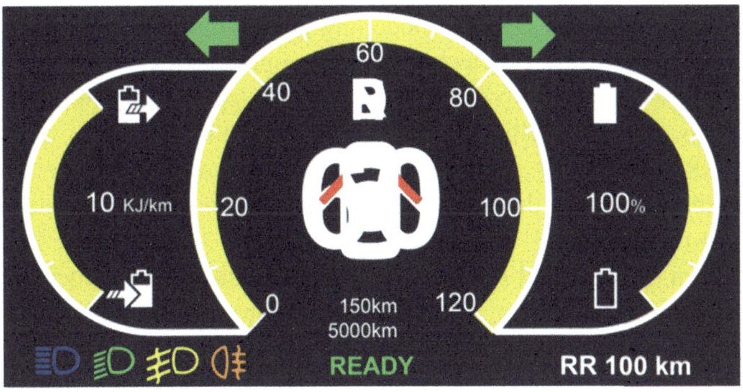

Fig. 13 Example of display layout for the POLLUX demonstrator

To prevent driving with over speed, the AFFP will give the driver vibration feedback if the vehicle speed is higher than the setting of the maximum speed of the pedal. In case that the driver accelerates the vehicle too quickly, the AFFP will also give the driver a vibration feed-back until the acceleration becomes lower than a predefined acceleration ratio.

The evaluation and testing of the demonstrator have been done on real routes, City route and Duracar New European Driving Cycle (NEDC) route. City route consists of 4× Urban Driving Cycles (UDC) and Duracar NEDC route consists of 4× UDC cycles and 1× Extra-Urban driving cycle (EUDC) with the maximum allowed road speed of 120 km/h and for low-powered vehicle 90 km/h.

Recommendation after the evaluation test with the QUICC DiVa demonstrator is to link the maximum speed of the pedal to the maximum road speed. It is believed that the energy saving ratio will be significantly improved.

7 Summary and Conclusions

The EV architectures will change the overall domain partitioning and their boundary conditions together with the distribution of the core system functions and more functions will be performed by distributed embedded systems. Communication protocols will play a crucial role in that high speed protocols (i.e. FlexRay) will be required for a safe and reliable management of the architecture.

The EV architecture layering and the introduction of the concept both on the level of the EV (electronics, communication, energy management, software layers) and as well on the individual layers in the architecture (i.e. communication layer with Ethernet backbone) becomes a necessity.

The concept in case of the communication layer opens the way of implementing the future Internet of Vehicles (IoV) paradigm.

In this context the key technological innovations achieved by the ARTEMIS POLLUX project are listed below:

- Development of a platform concept for EV architecture, electronics, communication and embedded systems modelling and simulation.
- Layered EV architecture (electronics, communication, energy management, software) based on ICT models. Smart domain partitioning into autonomous sub-systems across a multi-layer domain abstraction.
- AUTOSAR® compliant parallel computing real-time and secure platform to optimise the computing devices while maintaining the fault tolerance, safety and enhancing reliability.
- Specification, design, and development of two multi core microcontroller platforms validated for motor controller multi propulsion system and battery management modules.
- Wired and wireless deterministic, safe, secure and reliable links and power-optimised networking technologies enhanced towards EV constraints

(determinism, bandwidth EMI/ESD/EMC compatibility). Implementation of novel circuits for in-vehicle network links (Ethernet validation in EVs, CAN Partial Networking and FlexRay networks)

  - Power saving and cost saving physical layer devices for CAN networks
  - FlexRay devices with improved EMC capabilities addressing the need of EV environment and simulation platform for the network simulation in EVs
  - Data-time flow simulation environment for CAN, FlexRay and Ethernet networks which enables network load validation and optimisation of networking structures
  - Reliable wireless communication environment and models for remote monitoring systems and wireless backup for non-critical sensor data transmission

- Ethernet network architecture as EV backbone communication environment
- X-by-Wire fast prototyping technology and novel steer-by-Wire technology with torque actuation. X-by-Wire concept for semi-automatic parking assistance for urban EVs.
- Novel HMI using state-of-the-art display technologies applied in the environment of EVs and customizable haptic feedback system for EV steer-by-Wire interfaces.
- Reference programming models and multi-thread Real-Time Operating Systems. HW-SW integration and control safety.
- Novel HW/SW solutions for Li-ion cells and super capacitors monitoring and energy balancing. Vehicle to grid (V2G) interface and integration with battery management systems.
- V2G multisystem including energy management system. V2G interface and battery management systems integration.
- Battery and super capacitors management systems addressing the requirements of the EV type and range.
- Scalable centralized powertrain computer managing the propulsion environment of the EV and enabling fast prototyping.
- Motor control functionality for centralized and decentralised propulsion architectures applying the ISO 26262 safety concept.
- Multi core microcontroller platforms for the specific usage of EV computing requirements for motor control and propulsion systems.
- Remote monitoring platform for EVs where all communication, sensors and other important parameters of the EV can be monitored through a diagnostic interface.
- Successful integration, validation, verification and testing at test bench and into 5 EVs demonstrators (Fig. 14).

Duracar
HMI X-by-Wire, Force Feedback Pedal

Think
X-by-Wire, Parking Assistance

Fiat 500 HEV
Multicore processing motor controller, FlexRay Networking, Battery Management System

City Motion
Vehicle Controller, HMI, DC-DC Converter, Battery Management System, Partial CAN Networking

Test Bench Vehicle - PSA Peugeot Citroën
Battery Management, Multi propulsion power train, Multicore processing, PLC networking, FlexRay networking, V2I, V2G integration

Fig. 14 ARTEMIS POLLUX EVs demonstrators

Acknowledgments The work was supported by the ARTEMIS JU POLLUX project under the project grant. The authors would also like to convey thanks to the Public Authorities and European Commission for providing the financial support. The authors wish to acknowledge the fruitful discussions with the members of ENIAC JU E^3Car and Internet of Energy for Electric Mobility (IoE) projects.

References

1. Chan CC, Bouscayrol A, Chen K (2010) Electric, hybrid and fuel- cell vehicles: architectures and modeling. IEEE Trans Veh Tech 59(2):589–598
2. ISO/DIS 26262-1 (2009) Road vehicles—Functional safety, July 2009
3. Infineon Technologies AF (2012) TriCore™ V1.6 User Manual, 2012
4. ISO 15118 Road vehicles—vehicle to grid communication interface
5. Eckert C (2010) IP-basierte Kommunikationen im/zum Fahrzeug, SafeTRANS Workshop Presentation, p 6
6. http://www.avnu.org
7. Plankensteiner M (2011) Ethernet learns to drive Hanser automotive. TTTech Viena 12:14
8. Hank P, Müller S, Vermesan O, Van Den Keybus J (2013) Automotive ethernet: in-vehicle networking and smart mobility. In: Proceedings design, automation and test in europe conference and exhibition (DATE), pp 1735–1739. ISBN:978-1-4673-5071-6
9. BroadR-Reach Specification OPEN Alliance. http://www.opensig.org
10. Vermesan O, Blystad L-C, John R, Hank P, Bahr R, Moscatelli A (2013) Smart, connected and mobile: architecting future electric mobility ecosystems. In: Design, automation & test in Europe conference & exhibition (DATE), pp 1740–1744. ISBN:978-1-4673-5071-6
11. Miller JM, Startorelli G (2010) Battery and ultracapacitor combinations-where should the converter go? In: IEEE vehicle power and propulsion conference (VPPC), pp 1–7
12. Intelligent Parking Assist System. http://en.wikipedia.org/wiki/Intelligent_Parking_Assist_System
13. Mercedes Parktronic. http://techcenter.mercedes-benz.com/en/active_parking/detail.html
14. Active Park Assist FORD, www.youtube.com/watch?v=xW-MhoLImqg
15. Park Assits BMW, www.youtube.com/watch?v=v7DxzsbAZxs

End-to-End Integration of the V2G Interface with Smart Metering Systems (Results of the EU Co-funded FP7 Project "PowerUp")

Andras Kovacs, Robert Schmidt, Dave Marples and Raduz Morsztyn

Abstract Standardization efforts are currently underway to realize the Europe-wide deployment of Smart-Grids. Numerous protocols have already been standardized; each tailored to a distinct application. Amongst these, Smart Metering and EV charging have only recently started to converge, and the goal of the PowerUp project has been to advance such convergence. This paper presents the major results from the project. After introducing the underlying system architecture for the end-to-end integration between the Vehicle-to-Grid (V2G) communications interface and Smart Metering systems, we describe critical Smart-Grid integration aspects for each protocol layer within the V2G communications protocol stack. Finally, prototype test observations are presented. The outlined end-to-end integration of the V2G interface demonstrates the technological solution for ensuring that even mass-deployment of EVs would not interfere with the stability of the electric grid.

Keywords EV to Grid communication · V2G communication interface · Smart metering infrastructure · PLC technologies · Interoperability

A. Kovacs (✉)
BroadBit, Eotvosova 12, 94501 Komarno, Slovakia
e-mail: andras.kovacs@broadbit.com

R. Schmidt
DENSO AUTOMOTIVE Dtld, GmbH Freisinger Str. 21, 85386 Eching, Germany
e-mail: r.schmidt@denso-auto.de

D. Marples
Technolution, Zoudelijk Halfrond 1, 2801 DD Gouda, The Netherlands
e-mail: dave.marples@technolution.eu

R. Morsztyn
Corinex Communications, Klukata 6, 82105 Bratislava, Slovakia
e-mail: raduz.morsztyn@corinex.com

B. Müller and G. Meyer (eds.), *Electric Vehicle Systems Architecture and Standardization Needs*, Lecture Notes in Mobility,
DOI 10.1007/978-3-319-13656-1_9

1 Introduction

Preparing for the mass market deployment of EVs requires the development of an Vehicle to Grid (V2G) communication system for managing the EV recharging sessions to allow grid balancing and the correlated demands causing overloads. The standardization of an interoperable V2G communications interface is ongoing in the ISO/IEC 15118 working group [1]. This standard defines the communications interface between the EV and the local recharging controller device. On the automotive side of the V2G interface proprietary integration with the battery management and HMI components, usually over CAN or a similar bus structure, integrates the recharging functionality into the vehicle. On the grid side, integration with Smart Metering technologies under current deployment is expected to be the key for a successful introduction of an EV to Grid communication system. These integrations, on the vehicle and grid sides, are the focus of PowerUp, and this article.

In PowerUp an end-to-end EV to Grid communications system has been developed, implementing all the above integration aspects of the V2G interface [2]. On the automotive side of the V2G interface this system integration with BMS and HMI, while on the grid side of the V2G interface the solution achieves integration with Smart-Metering infrastructure. This resulting end-to-end EV recharging control system has been validated under both AC and DC charging scenarios. Figure 1 shows the system architecture view of the implemented solution. From an architectural perspective, ISO/IEC 15118 defines the 'Supply Equipment Communications Controller' (SECC) infrastructure-side entity for the V2G interface, while Smart-Grid systems terminate at the 'Smart-Meter' (SM) entity. The key for successful integration and deployment is therefore to develop the right interaction concept between these SECC and SM entities, which are both part of the local EV Supply Equipment (EVSE). The next section addresses physical deployment issues of these entities, and subsequent sections describe layer-by-layer integration issues of the involved communication protocol stacks.

2 Physical Deployment of the V2G Infrastructure

2.1 Deployment for Private Recharging Stations

Consequent to the EU Smart Metering directive [3], nearly all EU households shall be equipped with a SM over the next 5 or 6 years. Most Smart Meters have remotely upgradeable firmware, and could thus conceivably be upgraded to incorporate a SECC entity, subject to CPU and memory constraints. While currently deployed SMs are mostly using narrowband S-FSK PLC technology, the next wave of SMs is expected to exploit OFDM-modulated PLC technologies. Dual-mode G3/PRIME chipsets under development, as well as the superior capabilities of such OFDM links in comparison with S-FSK, mean that SMs are

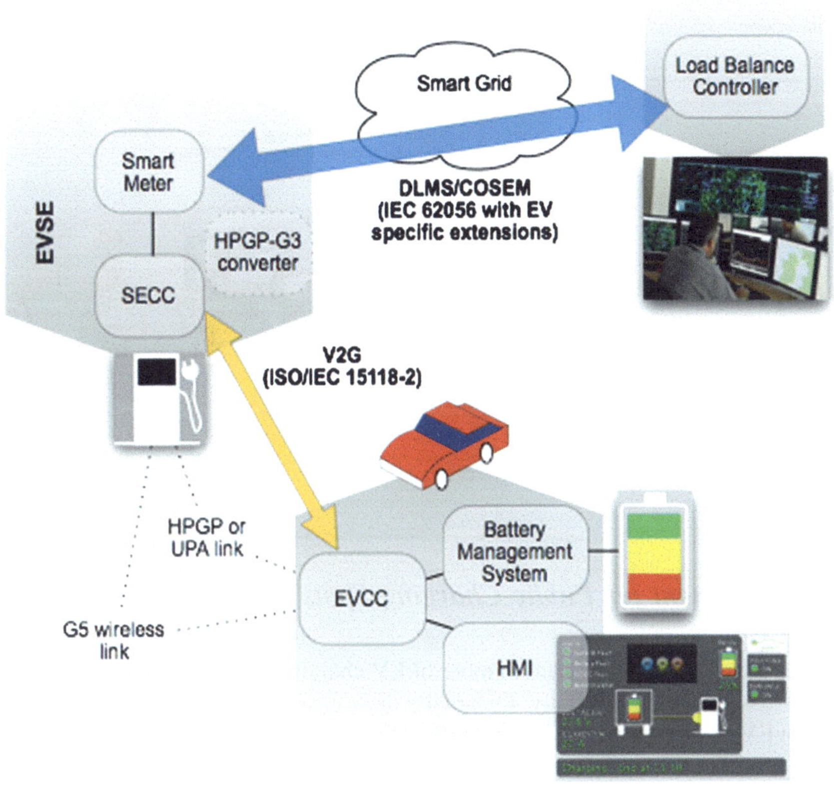

Fig. 1 The EV to Grid communications system architecture in PowerUp

expected to converge towards G3/PRIME or a subsequent standards-based PLC communication model. However, ISO/IEC 15118-3 specifies the 'HomePlug GreenPhy' (HPGP) technology for the PLC link in the V2G interface and so an issue of physical layer compatibility and interoperability arises and cost constraints make it unlikely that SMs include any EV-specific additional hardware. A solution to this problem would be to use a bridge between different media in the wall-box outlet installed for the EV power cable. This wallbox would then bridge the SM-specific and EV-specific PLC links. Such a HPGP-to-G3 media bridge has been prototyped in PowerUp, and is shown in Fig. 2. Sufficient miniaturization has been already achieved, and its size can be easily reduced further in subsequent product development for incorporation into the envisaged wall-box outlet.

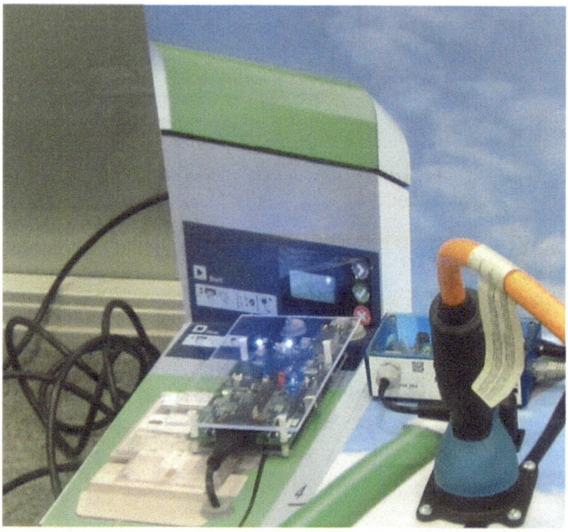

Fig. 2 HPGP-to-G3 PLC media bridge

2.2 Deployment for Public Charging Spots

The deployment of a significant number of EV charging spots is already underway both across the EU and globally. Eventually these charging spots must be retrofitted with an EV to Grid communication system to allow them to participate in the smart grid model. The PowerUp project has demonstrated the compact implementation of the additional components required for such a migration, which can be fitted into commercial EV charging spots. Figure 3 shows an example for such retrofitting.

2.3 Deployment for Public Transport and Goods Delivery Vehicles

With their demand for high-power fast recharging, public transport vehicles present special requirements for the V2G interface. Moreover, their physical connection may be constrained in the number of conductors available, with no capacity for the additional 'Pilot Control Lines' that are assumed in most of the current standards. This would typically be the case for a pantograph connection, for example. This necessitates a deviation from the ISO/IEC 15118-3 standard, which mandates a dedicated 'Pilot Control Line' for the V2G link and connection status signaling. For this environment, the V2G communications must be carried over the power line

Fig. 3 Retrofitting of a public recharging spot for EV to Grid communications

Fig. 4 The validation of the V2G interface coupled to the mains connection

connection, and a reliable means must be found for monitoring the connection status of the powerline. These deployment requirements have also been met by the prototypes developed in PowerUp. As shown in Fig. 4, a V2G interface connection over mains has been developed and tested with an electric bus. The following section gives details of the link control methodology and observed performance.

3 Link Layer Integration of the V2G Interface

3.1 PLC Link Performance

A QCA7000 chip based HPGP modem has been developed for establishing the ISO/IEC 15118-3 compliant PLC link. This device has been used with the default link control settings, along with a Marvell 9500 series chipset based UPA modem which has been used for analysis and comparison of HPGP to the established broadband over powerline Smart Grid system.

The PLC link performance has been tested over the powerline connection with both AC and DC charging systems. The alternative would have been to test on a dedicated 'control pilot line', as envisaged in the current ISO/IEC 15118 draft standards; however such alternative would not meet the deployment requirements described in Sect. 2.3. Since the noise environment is more challenging in the investigated setup, a stable solution under this setup is also stable in the alternative 'control pilot line' based PLC setup.

The PLC link quality has been tested by using the diagnostics of the UPA modem. Figure 5 summarizes these measurement results, which show that AC charging generates a higher noise on the powerline than DC charging. We measured an average SNR of 16 dB during 7 kW AC charging. As shown in Fig. 6, such powerline channel noise is well tolerated by the HPGP link, which maintains 4 Mb/s throughput regardless of the charging state. We have also measured the UPA system to achieve 20 Mb/s throughput under the same conditions. Moreover, the PLC throughput does not depend on the line coupling type; permitting both inductive and capacitive coupling solutions according to other deployment constraints.

Initially, a stability issue in the PLC link was observed. This correlated to the dynamics of channel parameters during the changes of mains charging state (i.e. between standby and charging states). During these state changes the PLC system has to rapidly adjust its modulation and bit loading to retain the communication link. In this situation link negotiation requires very quick assessment algorithms, whereas ensuring stability requires conservative and robust evaluation. Since link evaluation algorithms have common inputs to the link retention algorithms, the UPA system was tuned to achieve both fast link setup and stable communication link. This was made possible by creating configurable algorithm parameters that were adjusted based on the identified charging system state change signatures. This lead to improved link stability which was experimentally verified. The conclusion from these tests is that a PLC link for the V2G interface is feasible on the powerline connection, and it provides comparable bandwidth to a 'Pilot Control Line' dedicated for signaling. Achieving a stable link under all circumstances does, however, require that the PLC link parameters are tuned to the characteristics of the deployment environment.

An observed issue with the default HPGP link control settings has been the delay of link establishment. It took up to 30 s to establish an HPGP link after plug-in, which is too slow for some applications, including public transport charging.

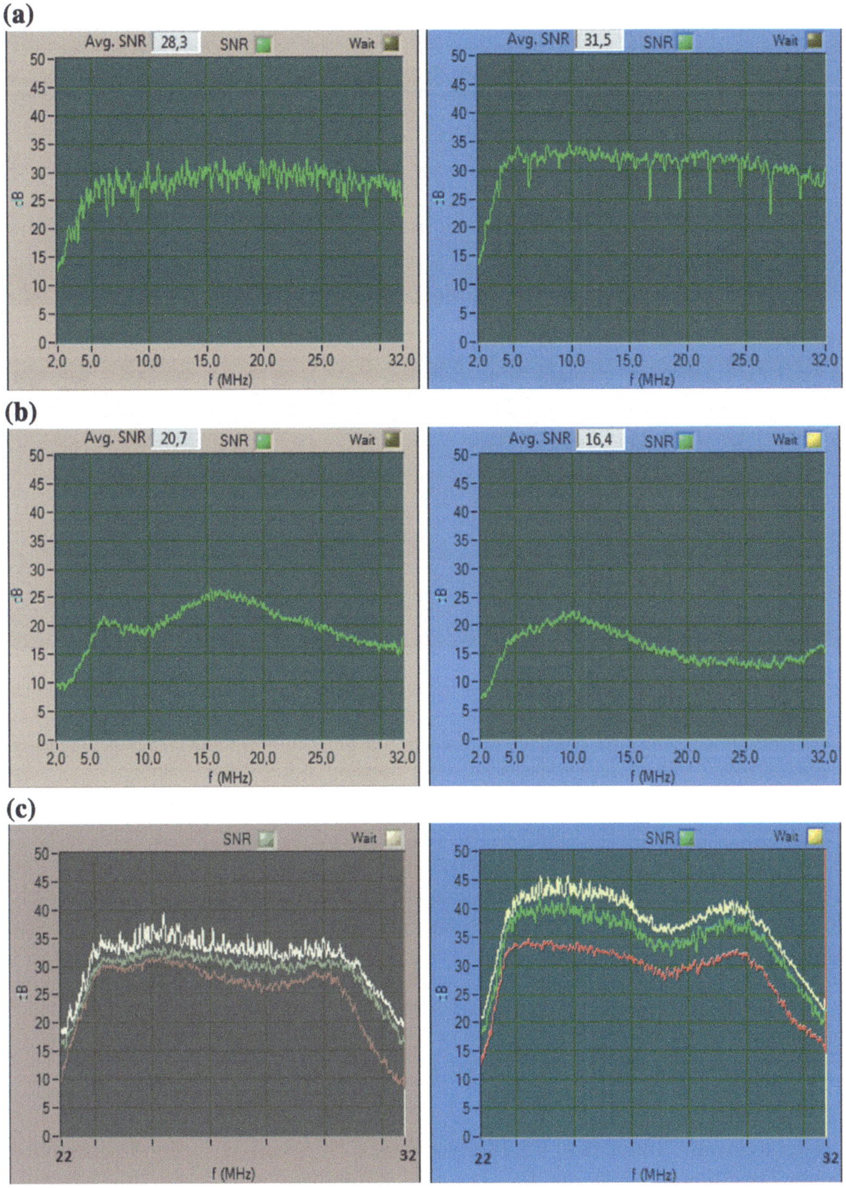

Fig. 5 The SNR ratio of the broadband PLC spectrum for the infrastructure-side modem (*pink background*) and the EV-side modem (*blue background*). The *horizontal axis* represents frequency (MHz scale), while the *vertical axis* represents SNR ratio (dB scale). **a** No recharging. **b** 7 kW AC recharging. **c** 5 kW DC recharging (at 600 V)

(a)

(b)

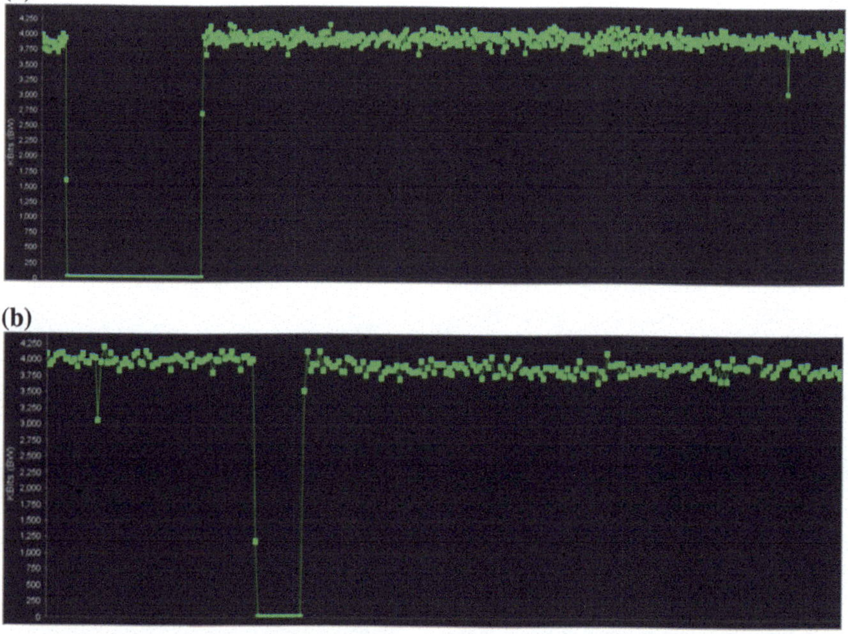

Fig. 6 HPGP throughput testing with inductive coupling (**a**) and capacitive coupling (**b**). The *horizontal axis* represents time (0–400 s scale), while the *vertical axis* represents TCP throughput (0–4.25 Mb/s scale). The three phases of the timeline are: plugged-in without charging, unplugged, plugged-in with 7 kW AC recharging. **a** Inductive coupling. **b** Capacitive coupling

We believe that it will be possible to perform similar HPGP link parameter tuning as per the UPA case above, but this was not verified in the project. Alternatively, some other PLC technology may be used when fast control link establishment is required.

3.2 Compatibility with Other PLC Media in the Smart-Grid

In order to ensure Smart Grid the PLC links used by Smart Metering devices much remain operational while the EV charging equipment is in use. This has been verified by measuring the throughput of a G3 PLC link, having one modem at the recharging spot, while the EV is recharging using AC technology. We measured 50 kb/s throughput on the G3 link regardless of the EV charging status. While this is a promising result, it must be noted that the link attenuation between the two G3 modems was very low due to the short distance between them modems, and that no conductive noise was imposed due to the isolated setup. A future follow-up test could be to measure the G3 link performance during EV recharging while attenuating the G3 link for emulating the proper distance of the remote side, and exposing it to the unfiltered grid. It is appropriate that this kind of testing is performed in a Field Operational Test (FOT) environment rather than the limited trial environment of the PowerUp project.

4 Network and Transport Layer Integration with the Grid Infrastructure

The networking topology of the end-to-end EV to Grid communications system can be described as a sequence of two connections:

(a) EV to SECC connection: This uses UDP transport with link-local IPv6 multicast addressing to distribute the SECC Discovery Protocol messages. It uses TLS with unicast IPv6 addressing to deliver the V2G application messages. Figure 7 summarizes this protocol stack.

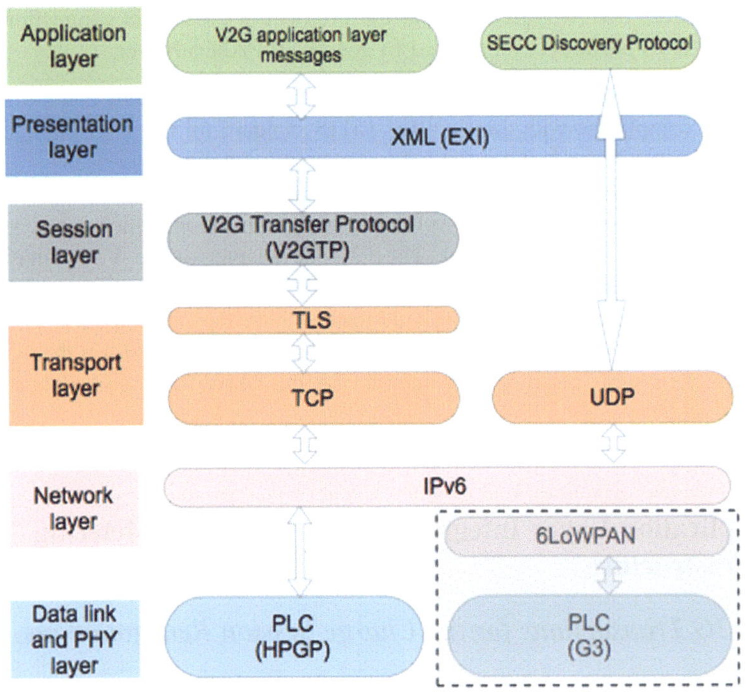

Fig. 7 The communications protocol stack between the EVCC and SECC. The *dashed-line* box represents IPv6-to-6LoWPAN bridge needed in case of HPGP-to-G3 media conversion, as potential extension to the ISO/IEC 15118 protocol stack

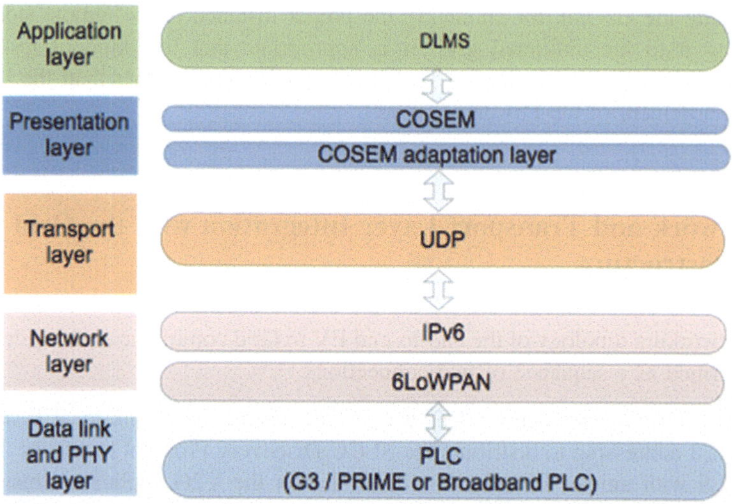

Fig. 8 The communications protocol stack between the SM and the Smart-Grid control center

(b) SM to Control Center connection: This is a point to point connection using UDP over IPv6. Figure 8 shows the protocol stack of this connection. The UDP/IPv6 header is compressed by the 6LoWPAN sublayer.

This network/transport topology is fairly straightforward, and can be implemented in compliance with the ISO/IEC 15118 standard for the V2G interface and the IEEE P1901.2 standard for the G3 based smart-metering interface.

Considering the anticipated private recharging scenario, there is one networking issue arising in case of the HPGP-to-G3 PLC link conversion, which is not covered by the ISO/IEC 15118-1 standard. The EV-side adapter of the V2G interface is sending IPv6 packets over the broadband PLC link, while the EVSE-side adapter of the V2G interface is sending 6LoWPAN packets over the G3 PLC link. Therefore the HPGP-to-G3 converter module in the wall-box must also perform the bridging between plain IPv6 networking protocol and the 6LoWPAN encapsulated networking protocol.

5 Application Layer Integration with the Smart Metering Infrastructure

5.1 V2G Transactions for the Charge Session Renegotiations

The ISO/IEC 15118 describes the communication interface, covering a whole protocol stack and its diverse use cases for establishing, maintaining, and terminating the EV charging session. In particular, the possibility of renegotiating charging schedules is a key feature for dynamic control over the EV's charging process, designed to maintain power grid stability. We distinguish two types of renegotiations:

(a) User-initiated renegotiation: The driver updates charging parameters, consisting of departure time and requested energy amount. For most use cases, only the planned departure time is altered. Typically, the requested energy amount is not altered, since users are assumed (but not constrained) to charge until the battery reaches fully charged status. To signal this desire, the EVCC can set a flag in the ChargeProgress data element for ReNegotiation. Both sides then know that the next messages will negotiate the new charge parameters before the charging will resume.
(b) Grid-initiated renegotiation: The grid controller updates the charging conditions (automatically or manually). Typically, this will lead to a change of the peak charging power in order to accommodate arising grid bottlenecks. Ideally, the desired departure time of the user won't be changed.
During the charging loop, the SECC has to signal that it desires a renegotiation, to be started by the EVCC. The SECC can set a flag for ReNegotiation in any message of the charge loop. Upon reception, the EVCC has to perform the same steps a described above.

Both of these renegotiation types have been implemented in PowerUp using the current ISO/IEC 15118-2 message set, i.e. without specifying new non-standard messages. The resulting V2G application layer thereby fully complies with the ISO/IEC 15118-2 standard, while also allowing dynamic adjustments to the charging process of the EV.

5.2 Interaction Between V2G and DLMS/COSEM

The key to successful application layer integration between the EVs and the electric grid is to establish a proper interaction between the V2G application layer communicating towards the EVs and the DLMS/COSEM application layer [4] communicating with the grid control center. DLMS/COSEM transactions generally consider the SM entity as a server, which the grid control center may query via the GET service or configure via the SET service. However a new PUSH service has been recently introduced, which allows the SM to initiate the passing of some information to the grid control center. Our implementation of the V2G-DLMS/COSEM interaction makes full use of this PUSH capability and the concept of this interaction is shown in Fig. 9. The SECC acts as a server on the V2G interface, meaning that any V2G transaction is initiated as a request to the SECC. During the setup phase of the V2G charging session each incoming request triggers a corresponding PUSH transaction to the grid controller, which then uses the

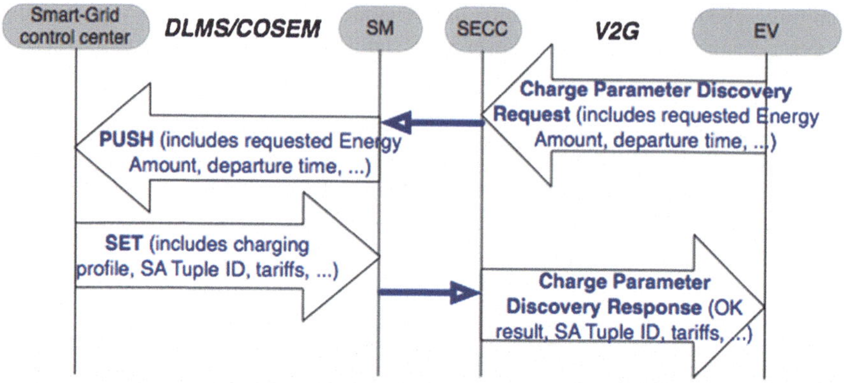

Fig. 9 Application layer interaction between the ISO/IEC 15118 V2G and IEC 62056 DLMS/COSEM interfaces during the charging session establishment

SET transaction to load the response parameters into the SM. Upon receiving these parameters, the SECC responds to the incoming V2G request. This mechanism is illustrated in Fig. 9 through the example of the 'Charge Parameter Discovery' transaction. Analogously, this same methodology is applied for the other remaining V2G transactions during the charge setup phase.

After starting the EV recharging session, V2G transactions comprise periodic 'Charging Status' and 'Metering Receipt' request/response messages. These requests can be locally handled by the SM via the SECC, without interacting with the grid controller. Therefore the grid controller can determine its own periodicity for querying the amount of energy transferred to the EVs (using GET transaction to the SM) or can instigate a renegotiation of the charging session if needed (using a GET transaction to the SM containing the same parameters as illustrated in Fig. 9). This solution is crucial for not exceeding the (limited) capacity of the PLC link of the SM, because this link is potentially shared among a large number of SMs. Only the setup of the V2G session requires additional messages, whereas the energy consumption of charging process can be monitored at the same rate as ordinary SM queries, and thus can be piggybacked.

The presently published DLMS/COSEM messages in the IEC 62056 standard do not contain EV-specific objects. The PowerUp projects has therefore specified the required DLMS/COSEM object extensions for performing the above-described interaction with the V2G interface, and has published these specifications on its project website [2]. These proposed extensions are scheduled for discussion at the DLMS User Association during September, 2013, and we anticipate them to become part of a future release of the IEC 62056 standards. Therefore we expect the PowerUp solution for application layer integration with the grid infrastructure to be compliant with upcoming releases of the relevant standards.

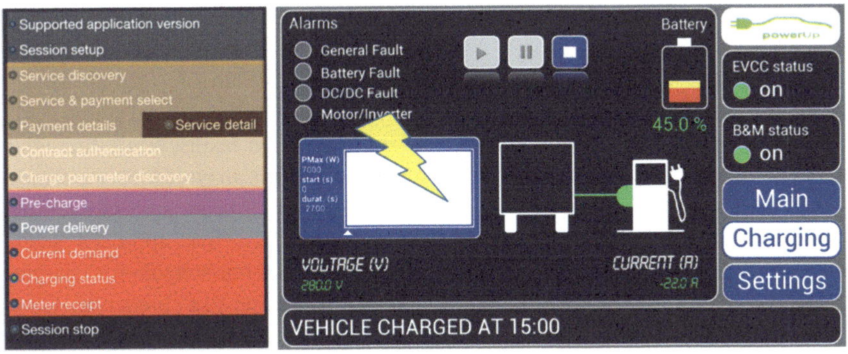

Fig. 10 User interfaces at EV-side. The screen on the *left* shows a diagnostic screen for monitoring the progress of the V2G session. The screen on the *right* shows the prototyped driver interface, displaying the charging schedule and status

For the user, this application layer integration is transparent. Those processes may be configured and supervised through intuitive user interfaces; Fig. 10 illustrates the prototype implementation of such user interface.

We believe that our choice of designing an interaction between the ISO/IEC 15118 V2G interface and the IEC 62056 Smart-Metering interface is well justified by their current industrial relevance:

- ISO/IEC 15118 represents a global consensus for the V2G interface between the EV and EVSE entities, with experts from EU, USA, and Japan/China/S. Korea contributing to the development of this standard.
- IEC 62056 DLMS/COSEM represents a global consensus for the Smart-Metering interface; this standard has been adopted for Smart-Metering systems deployment nearly everywhere in the world [5] (with the exception of the USA and Germany).
- Smart-Metering systems are expected to be widely deployed. In PowerUp, it was proposed to combine Smart-Metering and SECC functionality within one entity, as a suitable deployment solution for private EV charging scenarios. As for public EV charging scenarios, a similar solution could be applied, with additional mechanisms for identification and authentication.

The above observations imply that the selected PowerUp solution for application layer integration with the grid infrastructure makes the project results globally relevant and easily deployable in the future.

5.3 Alternative Approaches

While the V2G-DLMS/COSEM interaction mapping described above represents one proposed solution, it is nevertheless not the only possible mapping. This subsection therefore briefly presents an overview of alternative approaches, pursued by other research groups.

One alternative is to use the Open Charge Point Protocol (OCPP) for the application layer messaging to the grid control center. OCPP is based on an XML information schema, with SOAP-over-HTTP presentation layer. While the availability of plentiful development tools for these protocols may reduce the implementation effort, the verbose nature of XML/SOAP/HTTP presents a problem for bandwidth-limited links, if uncompressed. Consequently, this approach seems applicable only if sufficient bandwidth is available, and is not suitable for charging spots relying on Smart-Grid networks established by shared PLC links over G3 or PRIME technologies.

Another alternative is the use of the IEC 61850 'Distributed Energy Resources' (DER) protocol for application layer messaging to the grid control center. The IEC 61850 interface has been designed for managing energy flows between renewable electricity resources and power users. While the presently published standard does not contain EV-specific objects, a proposed EV-specific extension has been published in IEC Technical Report 61850-90-8, which enables interaction with the ISO/IEC 15118 based V2G interface. This approach represents therefore a realistic alternative to the V2G-DLMS/COSEM interaction. In the opinion of the authors, the main drawback of this approach is the increased complexity of having to implement a triple system in the EVSE—consisting of V2G protocol stack for interaction with the EV, DLMS/COSEM protocol stack for Smart-Metering, and DER protocol stack for interaction with the Smart-Grid controller—as opposed to just V2G and EV-extended DLMS/COSEM entities.

6 Conclusions

The PowerUp research project results presented in this paper provide a suitable blueprint for the integration of V2G-based EV recharging with Smart Metering systems. The key project results entail the development and demonstration of an end-to-end EV recharging control system; on the automotive side of the V2G interface this involves integration with BMS and HMI, while on the grid side of the V2G interface this involves integration with Smart-Metering infrastructure. The presented solution is compliant with relevant communications standards, and has been designed for meeting all foreseen deployment scenarios: private re-charging, public re-charging spots, and control of public transport vehicle re-charging. A layer-by-layer investigation of the arising communications issues has been performed during the design and testing phases of the project. The resulting system implementation has shown a good end-to-end communications performance in the laboratory tests, and is therefore ready for field operational trials upcoming in the near future.

References

1. ISO TC22 and IEC TC69: ISO/IEC 15118 Road vehicles—Vehicle-to-Grid communication interface
2. www.power-up.org
3. EU Directive 2009/72/EC
4. IEC TC13: IEC 62056 Electricity metering—data exchange for meter reading, tariff and load control
5. www.dlms.com/organization/listofmembers/index.html

The manufacturer's authorised representative in the EU is Springer Nature Customer Service Centre GmbH, Europaplatz 3, 69115 Heidelberg, Germany. If you have any concerns regarding our products, please contact ProductSafety@springernature.com

Printed and bound by CPI Group (UK) Ltd, Croydon, CR0 4YY
01/10/2026
02225989-0007